Springer-Verlag 6900 Heidelberg 1 · Postfach 1780
Telefon (06221) 49101 · Telex 04-61723
1000 Berlin 33 · Heidelberger Platz 3
Telefon (0311) 822001 · Telex 01-83319

Springer-Verlag New York, NY 10010 · 175, Fifth Avenue
New York Inc. Telefon 673-2660

26 Fortschritte de chemischen Fereehung
Topics in Current Chemistry

Inorganic and Analytical Chemistry

Springer-Verlag
Berlin Heidelberg GmbH 1972

ISBN 978-3-540-05589-1   ISBN 978-3-540-36907-3 (eBook)
DOI 10.1007/978-3-540-36907-3

Library of Congress Catalog Card Number 51-5497.

Contents

The Dihalides of Group IVB Elements

Prof. John L. Margrave, Dr. Kenneth G. Sharp, and Dr. Paul W. Wilson

Department of Chemistry, Rice University, Houston, Texas, USA

Contents

J. L. Margrave, K. G. Sharp, and P. W. Wilson

Introduction

The elements of Group IVB (C, Si, Ge, Sn, and Pb) all possess the ground state configuration ns^2p^2. This configuration suggests that oxidation states of either two or four are especially probable for the elements in the group, and to a greater or lesser extent, both states are observed for each member. There is a continuous increase in the stability of the divalent state with respect to the tetravalent state with increasing atomic number, so that with carbon the (II)-oxidation state is restricted to the very reactive carbenes and to "special" compounds such as isonitriles, whereas (II) is the prevalent state in the inorganic compounds of tin and lead. Two factors probably contribute to this increase in stability of the divalent state. For one, the M-X bond energies generally decrease down the group, with the exception that the bonds between silicon and the most electronegative elements are usually stronger than the corresponding bonds to carbon [1]. Secondly, repulsive interactions between non-bonding and bonding electron pairs are larger for the smaller atom at the top of the group [2].

There has been some uncertainty as to the relative electronegativities of the various members of Group IVB. The current consensus seems to favor the order

$$C > Ge > Si \sim Sn > Pb,$$

but the difference in electronegativity between Si and Pb may be small [3]. At any rate, electronegativity differences seem inadequate to explain the monotonic increase of stability of the (II) state as one goes down the group.

The "inert pair" concept has sometimes been advanced to account for the extra stability of atoms or ions which contain a lone pair of s-electrons (e.g., Hg^{+1} or Tl^{+1}); however, this effect does not seem to be particularly operative in Group IV, since the 3rd ionization potentials are similar for the elements Si through Pb.

The chemistry of the dihalides of the Group IVB elements has developed along several lines. One approach has been to use the dihalides as reactive intermediates in liquid phase studies. For example, CCl_2 is produced by the alkaline hydrolysis of chloroform; this CCl_2 can then react with other reagents in the system. A very large amount of work has been done on this type of study and since it is already extensively described in the literature [4] it will only be briefly discussed in this article. The dihalides of Group IVB elements, particularly CF_2, are also intermediates in a number of gas phase reactions. Another important approach has been to design experiments that produce the dihalides in conditions that prevent their immediate reaction with other reagents in the system. This has permitted the direct measurement of some of their physical properties and also the determination of some of their descriptive chemistry. It has been principally this technique that has been used in the chemistry department of Rice University, Houston. In this article some of the results obtained using all of these approaches are described although the last technique will be emphasized.

A. Difluoromethylene

CF_2 is unique among carbenes because of its high stability and low reactivity. Investigations of the ultraviolet absorption spectrum of CF_2 have led to estimates of roughly 10 milliseconds to one minute for the half-life of CF_2 at pressures in the region of one atmosphere. The gas phase molecule does not react with BF_3, N_2O, SO_2, CS_2 or CF_3I at 120 ^oC[5]. The nature of CF_2 is perhaps best presented in separate sections discussing its preparation, structure and physical properties, reaction chemistry, and reaction kinetics.

Preparation. The majority of the preparations of CF_2 reported in the literature involve photolytic or pyrolytic processes. Table 1 contains a representative list of methods used to produce CF_2. Most of the reactions produce the molecule in its singlet ground state, but the reaction of O (^3P) atoms with C_2F_4 and the decomposition of CF_4 in a glow discharge appear to produce triplet CF_2. In this connection it is interesting to note that the reaction of Hg (^3P) atoms with C_2F_4 did not give rise to triplet CF_2; the authors suggested that the triplet C_2F_4 initially formed passes through an excited singlet prior to dissociation.

Structure. The ultraviolet emission spectrum of CF_2 was first examined by Venkateswarlu[22], who prepared the molecule by passing an uncondensed transformer discharge through CF_4. An extensive band system between 3250 and 2400 Å was observed. The similarity of the band system to that of NO_2 suggested that a non-linear triatomic molecule was responsible for the spectrum. Venkateswarlu identified the band system with the transition $^1B_2 \rightarrow {}^1A_1$.

The ultraviolet absorption bands were examined by Laird, Andrews and Barrow[23] who obtained much the same results as Venkateswarlu although they suggested that the band numbering previously assigned might be incorrect. They demonstrated that the observed system involves the ground state of CF_2. Since their equipment design permitted examination of only long-lived species, they estimated that the half-life of CF_2 is approximately 1 second at a pressure of 1 mm Hg.

More recent experiments[24] have resulted in the re-assignment of the band system origin, the extension of spectral measurements to shorter wavelengths, and the correlation of the observed absorption spectra solely with the bending modes of the two states involved. In addition, Simons[25] and Margrave[26] have suggested that the spectra are due to the transition $1_{B_1} \leftarrow 1_{A_1}$, rather than $1_{B_2} \leftarrow 1_{A_1}$, as originally proposed. Mathews[27] has analyzed the rotational fine structure of the band at 2540 Å, and obtained the following values for molecular parameters:

$$\text{upper state,} \angle FCF = 134.8\,^o, r_{C\text{-}F} = 1.30 \text{ Å}\cdot$$

$$\text{lower state,} \angle FCF = 104.9^\circ, \; r_{C-F} = 1.30 \text{ Å}.$$

3

Table 1. *Methods of production of CF_2*

Starting Material	Products	Multiplicity	Ref.
Photolytic methods			
$CF_2=CF_2 \xrightarrow{h\nu}$	$2\,CF_2$	Singlet	6)
$CF_2=CF_2 \xrightarrow{h\nu}$ (Hg-sensitized)	$2\,CF_2$	Singlet	7)
$CF_2N_2 \xrightarrow{h\nu}$	$CF_2 + N_2$	Singlet	8)
$O\,(^3P) + C_2F_4 \xrightarrow{h\nu}$	$CF_2 + CF_2O$	Triplet	9)
Fluorocarbons $\xrightarrow{h\nu}$	CF_2 + various compounds	Singlet	10)
$ClF_2COCClF_2 \xrightarrow{h\nu}$	$CF_2 + ClCOCF_2Cl$	Singlet	11)
Pyrolytic methods			
$CF_3COCF_3 \xrightarrow{600\,^oC}$	$CF_2 + CF_3COF$	Singlet	12)
$\underset{\diagdown O \diagup}{CF_2-CF_2} \xrightarrow{120\,^oC}$	$CF_2 + CF_2O$	Singlet	13)
$CF_3Sn(CH_3)_3 \xrightarrow{150\,^oC}$	$CF_2 + FSn(CH_3)_3$	Singlet	14)
$(CF_3)_3PF_2 \xrightarrow{120\,^oC}$	$CF_2 + (CF_3)_2PF_3$	Singlet	5)
$\underset{\underset{X_2}{C}\diagup}{CF_2-CX_2} \xrightarrow{160-200\,^oC}$	$CF_2 + CX_2=CX_2$	Singlet	15)
Fluorocarbons $\longrightarrow$	CF_2 + other products		10)
Other methods			
$CF_4 \xrightarrow{Glow\ discharge}$	CF_2 + several other species	Triplet	16)
$C + CF_4 \xrightarrow{2000\,^oC}$	CF_2	Singlet	17)
$CHF_3 \xrightarrow{Shock\ wave}$	$CF_2 + HF$	Unknown	18)
$CF_2=CF_2 \xrightarrow{Shock\ wave}$	CF_2	Unknown	19)
$\underset{CF_2-CF_2}{\overset{CF_2-CF_2}{\mid\quad\mid}} \xrightarrow{r.f.\ discharge}$	CF_2	Unknown	20)
$CH_2 + CF_2=CF_2 \longrightarrow$	$CF_2 + CH_2=CF_2$	Unknown	21)

The infrared spectrum of matrix-trapped CF_2 (produced by the photolysis of difluorodiazirine, CF_2N_2) has been examined [28]. The three fundamental vibrational frequencies were determined to be 668, 1102, and 1222 cm^{-1}. The intensities of the two stretching fundamentals were sufficiently strong to permit observation of the corresponding absorption of $^{13}CF_2$, from which the bond angle of CF_2 was calculated to be approximately 108 ^{o}. The gas-phase infrared spectrum of CF_2 has been observed by Pimentel and Herr [29]. Difluorodiazirine was flash-photolyzed and the infrared spectrum of the products was immediately taken with a rapid-scan infrared spectrometer. Absorptions due to CF_2 were seen but the resolution of the instrument was insufficient for determination of the symmetry of the absorptions. The half-life of CF_2 was estimated to be 2.5 m sec.

Powell and Lide [30] observed the microwave spectrum of CF_2 using a fast-flow microwave spectrometer. The CF_2 was prepared by passing a weak r.f. discharge through C_2F_3Cl, CF_4 or $(CF_3)_2CO$. The absence of fine structure and observable Zeeman shifts provided evidence that the CF_2 was in the singlet ground state. The bond angle was determined to be 104.9 ^{O} and the bond length to be 1.30 Å, in complete agreement with Mathewa' values (see Ref. [27]). For comparison, the C-F bond length in CF_4 is 1.317 ± 0.005 Å [31].

The Heat of Formation of CF_2. Several of experimental approaches have been used to determine ΔH_f^0 (CF_2). The most common technique involves mass spectrometric measurement of appearance potentials. The earlier appearance potential measurements indicated that ΔH_f^0 (CF_2) = -30 ± 10 kcal. mole^{-1} [32], but it now appears this value is too high. Margrave and co-workers [33] reported a mass spectrometric study of the C_2F_4/CF_2 equilibrium between $1127-1244$ ^{O}K. Both second and third-law determinations of the enthalpy of reaction for $C_2F_4 \rightarrow 2 CF_2$ were made, yielding -39.3 ± 3 kcal. mole^{-1} for $\Delta H_f^0{}_{298}$ (CF_2, g).

Two groups have studied the pyrolysis of CF_2HCl and have calculated ΔH_f^0 (CF_2) to be -43 and -39.1 kcal. mole^{-1}, respectively [34]. Shock waves were also used to study the formation of CF_2 from C_2F_4 and CHF_3; values of ΔH_f^0 (CF_2) = $-.39.7 \pm 3.0$ and -40.2 ± 4.0 kcal. mole^{-1} respectively [18,19] were obtained. Other methods that have been used to determine ΔH_f^0 (CF_2) include the pyrolysis of CF_4 on graphite [35] and the observation of predissociation in ultraviolet absorption spectra [36].

Reaction Chemistry of CF_2. The reactions of CF_2 that have been studied to date fall conveniently into two categories: reaction in solution and reaction in the gas phase. Recently, however, there have also been some investigations of the reactions of matrix-isolated CF_2. No attempt will be made in this article to recount the large number of investigations into solution-phase dihalocarbene chemistry; a brief summary of dihalocarbene solution chemistry will be given in the following section. The interested reader is directed to several reviews of this subject [4].

In solution, dihalocarbenes are often produced from the basic hydrolysis of haloforms:

$$CHXYZ + OH^- \longrightarrow CXYZ^- + H_2O$$

$$CXYZ^- \longrightarrow CXY + Z^-$$

Hine has shown that the relative ability of substituent halogens to enhance trihalo anion formation is

J. L. Margrave, K. G. Sharp, and P. W. Wilson

$$I \sim Br > Cl > F$$

and that halogens facilitate carbene formation in the order

$$F \gg Cl > Br > I.$$

This latter sequence has been attributed to the relative ability of the halogens to supply unshared pairs to the electron-deficient carbon atom, as represented by the hybrids shown below

$$|\overline{X} - \overline{C} - \overline{X}| \quad +X = \overline{C} - \overline{X}| \quad |\overline{X} - \overline{C} = \overline{X}+$$

In the case of difluoromethylene, Hine [4] suggested that formation of the carbene is so favored that dehydrohalogenation occurs in a concerted fashion, with no carbanion intermediate.

Several other methods of generating dihalocarbenes in solution have been reported; the most useful of these appears to be the *thermolysis of phenyltrihalomethyl mercury compounds* as reported by Seyferth and co-workers [37a], although other organometallic precursors have also been employed [37b].

$$\phi\text{-Hg-CF}_3 \longrightarrow \phi\text{-Hg-F} \cdot + :CF_2$$

$$\phi\text{-Hg-CCl}_2Br \longrightarrow \phi\text{-Hg-Br} + :CCl_2$$

The advantages of this method of carbene synthesis are that reaction can be carried out in neutral solution, and that reaction yields are often dramatically improved. Thus, although reactions of dihalocarbenes generally do not give rise to products corresponding to single bond insertion, Seyferth has reported insertion of phenyl (trihalomethyl) mercury-generated carbenes into

C-H, Si-H, Ge-H, O-H, B-C, Hg-X, Si-Hg, Ge-Hg and Sn-Sn bonds

Much of the literature regarding dihalocarbenes is concerned with reactions of CX_2 with olefinic substrates to give 1,1-dihalocyclopropane derivatives. These reactions occur with retention of stereospecificity, as expected for singlet carbenes. Dihalocarbenes also exhibit strong electrophilic behavior towards olefins, and will often not react with weakly nucleophilic species if stronger nucleophiles are present.

Gas-phase Reactions of CF_2. In the gas phase, CF_2 is remarkably unreactive as compared to CH_2. This situation has been dramatically demonstrated by Mahler [5,38], who did not observe reaction between CF_2 (as produced from the pyrolysis of (trifluoromethyl)fluorophosphoranes at 120°) and BF_3, H_2, CO, NF_3, CS_2, PF_3, SO_2, CF_3I or N_2O. Mahler did report the following reactions:

$$CF_2 + I_2 \longrightarrow CF_2I_2 + CF_2ICF_2I + I(CF_2)_3I + F_2C\overset{\displaystyle CF_2}{\underset{}{\diagdown\diagup}}$$

6

$CF_2 + HCl \longrightarrow CF_2HCl$

$CF_2 + Cl_2 \longrightarrow CF_2Cl_2$

$CF_2 + O_2 \longrightarrow COF_2$

$CF_2 + MoF_6/WF_6 \longrightarrow CF_4$ + reduced metal fluorides

$CF_2 + CF_3C{\equiv}CCF_3 \longrightarrow$ 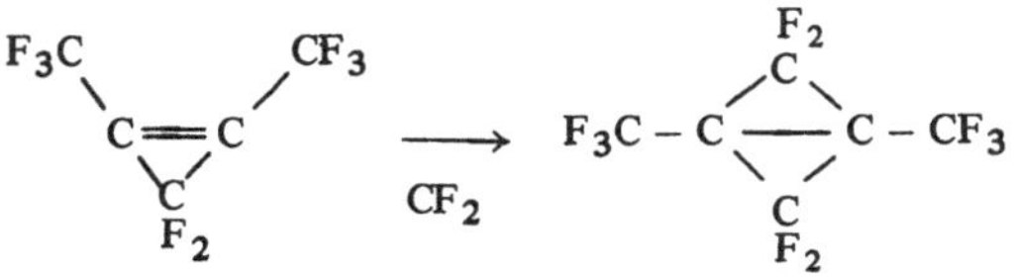

$CF_2 + F_3PO \longrightarrow CO + PF_5 + COF_2 + PF_3$

The CF_2/I_2 reaction is complicated by the fact that I_2 reacts with C_2F_4 to give $CF_2I\text{-}CF_2I$[39]. The HCl reaction is interesting since, as Mahler points out, it is the reverse of a reaction often used to produce CF_2. The reaction of CF_2 with O_2 is surprising in light of the many kinetic studies of the reactions of CF_2 in the presence of O_2, to be discussed later.

Mitsch [40] studied the reactions of CF_2 produced from the photolysis of CF_2N_2, and reported the following reactions

$CF_2 + Cl_2 \longrightarrow CF_2Cl_2$

$CF_2 + I_2 \longrightarrow CF_2I_2$

$CF_2 + N_2O_4 \longrightarrow$ 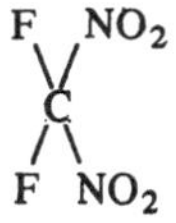

$CF_2 + NO_2Cl \longrightarrow$ 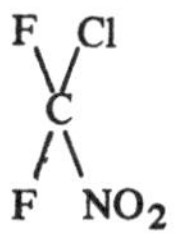

$CF_2 +$ 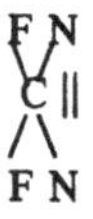 $\longrightarrow CF_2{=}N\text{-}N{=}CF_2 \xrightarrow{CF_2}$

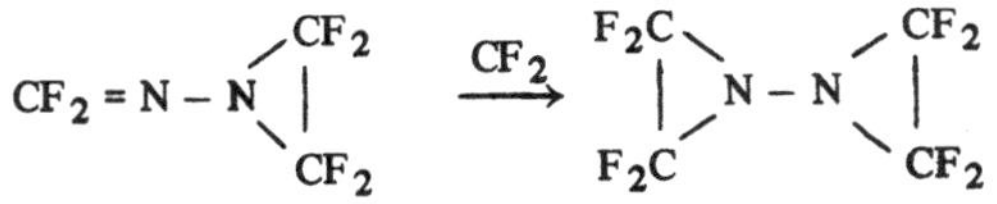

$$CF_2 + R\text{-}\overset{\overset{\textstyle O}{\|}}{C}\text{-}OH \longrightarrow R\text{-}\overset{\overset{\textstyle O}{\|}}{C}\text{-}OCF_2H$$

$$CF_2 + R\text{-}OH \longrightarrow R \cdot O \cdot CF_2H$$

$$CF_2 + RSO_3H \longrightarrow RSO_3CF_2H$$

$$CF_2 + CF_2\text{=}CF\text{-}CF\text{=}CF_2 \longrightarrow CF_2\text{=}CF\text{-}CF\overset{\diagup CF_2}{\underset{\diagdown CF_2}{\big|}} \longrightarrow CF_2 \overset{\diagup CF = CF}{\underset{\diagdown CF_2}{\diagup}} CF_2$$

Perfluoro-1,4-pentadiene and perfluoropropene undergo similar reactions. Atkinson and McKeagen [41] reported two similar reactions:

$$CF_3\text{-}CF\text{=}CF_2 + CF_2 \longrightarrow CF_3 - CF\overset{\diagup CF_2}{\underset{\diagdown CF_2}{\big|}}$$

$$C_2H_4 + CF_2 \longrightarrow H_2C \underset{CF_2}{\diagdown\diagup} CH_2 \quad \text{low yield}$$

Behind a shock wave CF_2 reacts with NO [42]:

$$CF_2 + NO \rightleftharpoons CF_2NO$$

$$2\,CF_2NO \longrightarrow 2\,CF_2O + N_2$$

$$CF_2NO + NO \longrightarrow CF_2O + N_2O$$

CF_2 also reacts with NOF. In this case the CF_2 used was prepared by the photolysis of C_2F_4 and the reaction was complicated by the reaction between C_2F_4 and NOF [43].

$$C_2F_4 + NOF \xrightarrow{hv} \underset{CF_2\text{-}CF_2}{F\text{-}N\text{-}O} \xrightarrow{CF_2} \underset{CF_2\text{-}CF_2}{CF_3\text{-}N\text{-}O}$$

Mastrangelo [20] has reported some interesting work on the *reactions of CF_2 trapped in matrices*. A stream of octafluorocyclobutane was passed through a radio frequency discharge and condensed on a liquid nitrogen-cooled cold finger. The resultant deposit was an intense dark blue which persisted until the cold finger warmed to ca. 95 °K. When radical generation times exceeded 15 minutes, however, the blue condensate slowly changed to a red color believed to be associated with $CF_3 \cdot$ radicals. On warming, the blue condensate gave rise to C_2F_4 and unreacted $c\text{-}C_4F_8$, but no polymeric residue. When chlorine was condensed on the blue deposit before warmup, CF_2Cl_2 and $CF_2Cl\text{-}CF_2Cl$, with smaller amounts of CF_3Cl, were observed in the products. Mastrangelo attributed the blue color to the presence of CF_2 radicals, and the ensuing red color to the disproportionation of CF_2 to CF and CF_3. No determination of the spin state of either the gas-phase or condensed species was reported; in view of the intense color of the condensate, the absence of polymeric radical chains, and the proposed disproportionation of CF_2 to CF and CF_3, the presence of triplet CF_2 seems quite possible.

Milligan and Jacox [44] have recently reported an elegant synthesis of CF_2 in an *argon matrix*. Carbon atoms, produced from the photolysis of cyanogen azide, were allowed to react with molecular fluorine, and the presence of CF_2 was demonstrated from infrared spectra. Use of radiation effective in photolyzing F_2 produced CF_3 from the reaction of the CF_2 with atomic fluorine.

Kinetic Studies of the Gas-Phase Reactions of CF_2. As mentioned above, when gaseous CF_2 is produced in the presence of substances with which it does not react, the products obtained are tetrafluoroethylene and perfluorocyclopropane [5]. The decay of CF_2 was originally thought to follow zero-order kinetics (that is, removal of CF_2 by means of diffusion to the walls of the apparatus [45,23]). A study of the flash-photolysis of C_2F_4 by Dalby [6], however, showed that CF_2 decay follows second order kinetics, and a rate constant of 1.7×10^7 (liter/mole $\cdot$ sec) at 25 °C was determined for dimerization of CF_2 to $CF_2{=}CF_2$. Dalby further observed that the rate of disappearance of CF_2 was independent of the concentration of oxygen, C_2F_4 or C_2H_4 at pressures as high as 40 cm for the latter two. He was thus able to set an upper limit to the rate constant for the reaction of CF_2 with these molecules of approximately 10^4 liter/mole $\cdot$ sec.

Cohen and Heicklen [46] investigated the mercury-sensitized photolysis of C_2F_4 and were able to determine the rate constant for the reaction

$$CF_2 + C_2F_4 = c\text{-}C_3F_6$$

to be $k_{C_2F_4} = 6.4 \times 10^7 \exp(-7500/RT)$ or 4.5×10^3 liter/mole $\cdot$ sec at 25 °C.

The ratio of $k_{C_2F_4}$ to the rate constant for dimerization was also found: $k_{C_2F_4}/k^{1/2}$ dim. $= 395 \exp(-6700/RT)$ (liter/mole $\cdot$ sec)$^{1/2}$. This ratio has a value of approximately 5.6×10^{-3} at 25 °C. Although this method of CF_2 production apparently does not yield triplet CF_2, the molecule may be generated *via* the reaction of ground-state oxygen atoms (^3P) with C_2F_4 to yield $^3CF_2$ and CF_2O [9,47]. Triplet CF_2, like the singlet molecule, can add to C_2F_4 to form $c\text{-}C_3F_6$. Triplet CF_2 can also revert to singlet CF_2 through a bimolecular reaction involving an excited C_2F_4 intermediate:

$$2\ ^3CF_2 \longrightarrow C_2F_4{}^* \longrightarrow 2\ ^1CF_2$$

The self-annihilation reaction occurs much faster than addition to C_2F_4. If molecular oxygen is added to the system, $^3CF_2$ may then react with O_2 to give CF_2O_2. This reaction is slightly faster than the combination of $^3CF_2$. The CF_2O_2 radicals produced in the reaction with O_2 are removed *via*

$$2\,CF_2O_2 \longrightarrow 2\,CF_2O + O_2$$

The presence of triplet CF_2 was inferred from the fact that in this system all of the CF_2 species are scavenged by O_2 if the O_2 pressure is greater than 5 torr, coupled with previous observations that the rate of reaction of singlet CF_2 with O_2 is extremely slow [6,7,48].

Modica and LaGraff [19] have conducted a series of examinations of the production and kinetic aspects of the reactions of CF_2 in shock waves. C_2F_4, diluted 1:100 with argon, was shocked over the temperature range 1200–1800 °K. Ultraviolet absorption of the shocked mixture revealed that dissociation of the C_2F_4 to CF_2 was virtually complete within 1 μsec. The dissociation reaction was found to be second order,

$$^1/_2 d[CF_2]/dt = K_{Ar}[C_2F_4]\,[Ar],\ \text{with}\ K_{Ar} = 7.82 \times 10^{15}\,T^1/_2$$

$$\exp\,(\text{-}55690/RT)\ \text{cc/mole} \cdot \text{sec}$$

The value of ΔH_f° (CF_2) calculated from the measured heat of the above reaction agrees well with that obtained by other methods [33,34], and lends strength to the assumption that equilibrium conditions prevail in the system.

When oxygen was added to the C_2F_4/Ar mixture, no reaction with the O_2 was observed below 1400° K. At temperatures above 1700° K, however, the bimolecular oxidation of CF_2 to (initially) $CO + 2F + O$ was found to occur with

$$K_{ox} = 2.82 \times 10^{10}\,T^{1/2}\,\exp\,(\text{-}13280/RT)$$

At temperatures in the range 2600-3700° K CF_2 itself decomposes to $CF + F$, with equilibrium expressed by

$$\log K_c\,(\text{mole/cc}) = \frac{\text{-}103000 \pm 5700}{2.303\,RT} \text{-}0.41 \pm 0.11$$

B. Other Carbon Dihalides

Despite the large body of literature discussing the preparation and reaction chemistry of dichloromethylene in solution, very few reports of the isolation of the molecule have appeared. The technique of forming Group IV dihalides from the reduction of the tetrahalide with the metal has proved to be of great utility for production of SiX_2 and GeX_2, but has not been successful in the case of carbon.

Schmeisser and Schröter studied the reaction of CCl_4 with activated charcoal at 1300°, and originally [48a] reported isolation of CCl_2 itself as a mobile, volatile liquid boiling at -20° C. A subsequent publication [48b] retracted the claim, explaining that an equimolar mixture of dichloroacetylene and chlorine had comprised the "CCl_2." The paper further stated the CCl_4 was in fact undergoing a surface catalyzed pyrolysis rather than reaction with the charcoal. Carbon is known to catalyze the decomposition of CCl_4 to $C + 2Cl_2$ [49]. Schmeisser et al. obtained the following products from CCl_4 pyrolysis (yields in parentheses):

$$C\ (35);\ C_2Cl_2\ (20);\ C_2Cl_4\ (40);\ C_2Cl_6\ (5);\ C_4Cl_6\ (0.1);\ C_6Cl_6\ (0.1).$$

Dichloromethylene was presumably the precursor of the C_2Cl_4, although the latter compound could have resulted from disproportionation of C_2Cl_6 to Cl_2 and C_2Cl_4. Blanchard and LeGoff [50] studied the decomposition of CCl_4 on a tungsten ribbon in the temperature range 1300-2000 °K. The CCl_4 vapor, at a pressure of 10^{-5} mm Hg, was made to flow past the ribbon and directly into the ionization source of a mass spectrometer, which was then utilized to analyze the products. Between the temperatures of 1300 and 1600°, the major pyrolysis products were CCl_2 and Cl_2; between 1600 and 1900° CCl_2 and Cl prevailed. When a *carburized* tungsten ribbon was used virtually identical results were obtained, indicating that the reaction

$$C + CCl_4 = 2\ CCl_2$$

was not important under the existing conditions. The ionization potential of CCl_2 was determined to be 13.2 ± 0.2 eV, and the appearance potentials of the various $C\text{-}Cl_n^+$ ions were used to calculate approximate bond dissociation energies of the corresponding neutral species.

Three groups have recently claimed to have isolated CCl_2 in *low- temperature matrices and to have observed the molecule spectroscopically*. Milligan and Jacox [51] prepared CCl_2 in a manner analogous to that described for their matrix synthesis of CF_2. Carbon atoms formed *in situ* from the photolysis of N_3CN were allowed to react with Cl_2 in an argon or nitrogen matrix at 14° K. Subsequent to irradiation, two new bands at 721 and 748 cm^{-1} were observed in the infrared spectrum of the matrix. The relative intensity of the bands remained constant under varying conditions. The features disappeared rapidly on warmup of the matrix, with corresponding growth of bands assigned to CCl_4. Moreover, ^{13}C isotopic studies demonstrated that the compound in question contained only one carbon atom. The above observations were taken as evidence for the existence of CCl_2 as the species in question, and the bands at 721 and 748 cm^{-1} were assigned to the stretching fundamentals of the molecule. The bond angle for CCl_2 was estimated to lie in the range $90 - 110^\circ$. The authors also reported a weak band system between 4400 and 5600 Å, with a band spacing of 305 cm^{-1}, to be associated with CCl_2- containing matrices. By analogy with known electronic spectra of CF_2, the system was attrbuted to a transition from the singlet ground state to the first excited state, with an extensive progression in the upper state bending vibration.

Andrews [52] isolated CCl_2 in an argon matrix by means of the reaction of Li atoms with CCl_4. CCl_3 radicals are formed from the abstraction of a Cl atom from CCl_4 by Li, and CCl_2 is produced from the secondary reaction

$$Li + CCl_3 \cdot = LiCl + CCl_2$$

The loss of CCl_2 absorption on matrix warmup was accompanied by the growth of bands attributed to C_2Cl_4. A complete isotopic analysis of the CCl_2 spectra supported the assignment of the stretching fundamentals as $\nu_1 = 719 . 5$ and $\nu_3 = 745.7$ cm^{-1}, in excellent greement with the work of Milligan and Jacox. The weak ν_2 (bending) mode was not abserved. The bond angle of CCl_2 was estimated to be $100^{\circ} \pm 9^{\circ}$, which strongly indicates that the observed species is in the singlet electronic configuration. Stretching force constants were calculated, and F_{C-Cl} was found to be lower than the corresponding value for CCl_4—a fact which Andrews claims to be evidence for lack of significant pi - bonded contributions to the C-Cl bonds. This result is surprising since doubly-bonded resonance hybrids have long been invoked to explain the stability of dihalocarbenes.

About the same time as the publication of Milligan and Jacox' and Andrews' work, Steudel [53] claimed to have observed the *infrared spectrum* of CCl_2 condensed from the pyrolysis (or decomposition in a high-frequency discharge) of several C-Cl compounds. CCl_4, C_2Cl_6, $CHCl_3$, and $CSCl_2$ were passed individually through a furnace at 900 $^{\circ}$C, and immediately condensed on a KBr window at 83 $^{\circ}$K. In each instance, a broad band in the IR spectrum at 896 cm^{-1} was seen. The absorption diminished in intensity as the matrix was warmed, finally disappearing at 160—200 $^{\circ}$K. Since the pyrolysis products in each case included C_2Cl_4, CCl_2 was assumed to be the common intermediate in each reaction.

Although CCl_2 may well have been an intermediate in the pyrolytic reactions reported by Steudel, it seems clear that the molecule is not responsible for the observed band at 896 cm^{-1}. Andrews [54] has recently described the infrared spectrum of matrix-isolated CCl_3, and located one of the stretching modes (ν_3) at 898 cm^{-1}. Since the reactions discussed in Steudel's work all produce C_2Cl_6 as well as C_2Cl_4, he likely observed the CCl_3 radical.

The area of *gas-phase chemistry of dichloromethylene* is as yet largely unexplored. Haszeldine and co-workers [37b] have prepared CCl_2 from the pyrolysis of CCl_3SiCl_3 and CCl_3SiF_3, The CCl_2 thus produced was observed to react with ethylene and a number of butenes in 85—95% yield, and with C_2Cl_4 in 69—85% yield. Addition to *cis-* or *trans-*2-butene occured with retention of stereospecificity. No report of the dimerization of CCl_2 to C_2Cl_4 was given. These preparations belong to the general class of α -elimination reactions of trihaloalkyl organometallics, several of which were discussed in the section on CF_2. In a variation of this type of work Skell and Cholod [55] prepared CCl_2 in the gas phase by pyrolyzing $CHCl_3$ at 1400 $^{\circ}$K. This pyrolysis was carried out immediately above a solution of olefins and the CCl_2 reacted with these olefins to give dichlorocyclopropane derivatives. The authors argue that this confirms the fact that free CCl_2 is indeed the intermediate in α-elimination reactions.

Other than some solution chemistry very little indeed is known about CBr_2 and CI_2 or about mixed dihalocarbenes. Tyerman [56] has observed the band spectrum of CFCl between 3736–3466 Å. Its main feature is a progression of bands with an average spacing of 386 cm^{-1}. He also observed that, in contrast with CF_2, CFCl reacts with O_2 at room temperature.

$$CFCl + O_2 \longrightarrow CFClO + O$$

C. Silicon Difluoride

If the gaseous species resulting from passing SiF_4 over elemental silicon at 1100–1400 °C are condensed at temperatures below –80 °C and subsequently allowed to warm to room temperature, a waxy, tough white polymer of composition $(SiF_2)_n$ is obtained [57]. Mass spectrometric analyses of the gas phase products of the Si/SiF_4 reaction indicate that SiF_2 and SiF_4 account for over 99% of the species present, with the percentage of SiF_2 typically near 60% [58]. Gaseous silicon difluoride is extraordinarily stable compared to dihalocarbenes and other silicon dihalides. Its half-life at a pressure of 0.2 mm has been estimated to be 150 seconds [58]. Unlike other Group IV difluorides, SiF_2 shows no tendency to form gas phase dimers, and is essentially unaffected by the addition of many other gases (except for oxygen, which facilitates formation of Si-O-F polymers on the walls of the apparatus).

The low-temperature condensate of SiF_2 is a yellow-brown paramagnetic solid which remains unchanged when maintained at -196 °C. If, however, another substance is co-condensed with the SiF_2, the low-temperature species can be made to react—usually on warming. The reaction chemistry thus investigated has proved to be quite extensive, and likely represents the most comprehensive study of the low-temperature chemistry of a high-temperature molecule. Results of the various examinations of the physical and chemical properties of SiF_2, most of which have been conducted in this laboratory, will be discussed in the following sections.

1. Gas Phase Spectra

Ultraviolet Spectra. The first direct evidence for the existence of gas phase monomeric silicon difluoride resulted from observation of emission spectra of the molecule in an electric discharge through SiF_4 [59]. The emission band system was subsequently extended [60]; however, both of these investigations are now thought to have resulted in erroneous vibrational numberings. The ultraviolet absorption spectrum was reported by Khanna, Besenbruch and Margrave [61], who employed the "usual" preparative technique of reducing the tetrafluoride with the metal. They measured 28 absorption bands in the region between 2325 and 2130 Å. The most striking feature of the spectrum was the appearance of a series of bands with a periodicity of 252 cm^{-1}. This progression was correlated with the bending frequency of the excited state. As was the case for CF_2, no direct evidence for excitation of stretching frequencies was obtained. The vibrationless transition is thought to lie at 2266.4 Å, and is likely a $^1B_1 \leftarrow ^1A_1$ transition.

Microwave Spectrum. Rao *et al.* [62] were able to observe the microwave spectrum of SiF_2 by generating the molecule from the high-temperature Si/SiF_4 reaction and pumping the reaction mixture through an absorption cell. The Si-F bond distance and F-Si-F bond angle were calculated to be 1.591 Å and 100°59', respectively. The bond angle is smaller, and the bond distance longer than one might anticipate, suggesting that bonding involves mainly p^2 hybridization of the silicon orbitals.

Infrared Spectrum. The infrared spectrum of gaseous SiF_2 has been recorded from 1050 to 400 cm^{-1} [63]. Two absorption bands, centered at 855 and 872 cm^{-1}, were assigned to the symmetric (ν_1) and antisymmetric (ν_3) stretching modes, respectively. The assignment was rendered difficult because of the considerable overlap of the two bands. The fundamental bending frequency occurs below the instrumental range of the study, but a value of 345 cm^{-1} can be determined from the ultraviolet study. The vibrational frequencies were combined with data from a refined microwave study [64] and utilized to calculate force constants and revised thermodynamic functions.

Mass Spectrum. There have been two investigations of the mass spectrum of SiF_2. In one experiment [58] the gaseous mixture of silicon fluorides obtained after passing SiF_4 over a column of Si held at 1150 °C was passed into a 5 lt. bulb and thence into a mass spectrometer. Only SiF_4 and monomeric SiF_2 were observed; no polymeric species of SiF_2 were seen. By isolating the 5 lt. bulb containing the SiF_2 from the furnace and then monitoring the decay of SiF_2, it was estimated that SiF_2 has a half-life of 150 seconds. In a second investigation [65a] SiF_2 was produced by heating a mixture of Si and CaF_2 to about 1500 °K. From this study the following values were obtained:

$$\Delta H^o_{a,298} \; SiF_2, g = 12.33 \pm 0.2 \text{ eV and thence}$$

$$\Delta H^o_{f,298} \; SiF_2, g = -139 \pm 2 \text{ kcal mole}^{-1}$$

This value for the heat of formation of SiF_2 is not too close to that determined by a transpiration method, -148 ± 4 kcal mole^{-1} [65b]. The discrepancy probably arises from the interaction between SiF_4 and SiF_2 to form Si_xF_{2x+2} at the higher pressures [65c].

2. Studies of SiF$_2$ Condensate

Infrared Spectrum. Since the reaction chemistry of SiF$_2$ known to date occurs at low temperatures in the condensed phase rather than in the gas phase, it is naturally of interest to investigate the kow-temperature condensate formed from gaseous SiF$_2$. The first such investigation was conducted by Bassler, Timms, and Margrave [66], and involved recording the infrared spectrum of matrix-isolated SiF$_2$ between the temperatures of 20-40 $^\circ$K. Fig. 1 illustrates spectra obtained when a gas-phase SiF$_2$/SiF$_4$ mixture was condensed on a CsI window at 20 $^\circ$K and allowed to warm. One notes that the peak at 811 cm^{-1} disappears much faster than the rest of the spectrum as the matrix is warmed. Furthermore, when the furnace-to-window distance is increased to 10 feet, or when nitric oxide is co-condensed with the SiF$_2$, the peak is absent altogether. This behavior suggests that the species responsible for the absorption is more reactive than monomeric singlet SiF$_2$——perhaps triplet SiF$_2$, excited singlet SiF$_2$, or SiF$_3$. The second spectral feature evident on warmup is the appearance of two new bands at 830 and 892^{-1}. These absorptions first appear at about 35° K, grow to maximum intensity at 38° (at the expense of bands now known to be due to monomeric SiF$_2$), and disappear rapidly on further warming. When the matrix is warmed to 50 $^\circ$K, the spectrum consists of broad bands identical with those of thin layers of (SiF$_2$)$_n$ at room temperature. These facts, especially when viewed in conjunction with the chemical characteristics of SiF$_2$ condensates, lead to the conclusion that the new bands are due to SiF$_2$ dimer. The same study also examined some of the earlier SiF$_2$ chemistry by co-condensing potential reactants in the matrix. Most illuminating of these experiments was that involving BF$_3$. Previous work had shown that the reaction of SiF$_2$ with BF$_3$ (to be discussed in more detail later) leads to a series of compounds BF$_2$(SiF$_2$)$_n$F, with n at least two. When the BF$_3$/SiF$_2$ matrix was allowed to warm, a series of bands not associated with "pure" SiF$_2$ spectra appeared. The bands began to appear when those associated with (SiF$_2$)$_2$ reach a maximum. Moreover, the new absorption corresponded closely with those of the gas-phase spectrum of SiF$_3$SiF$_2$BF$_2$, the major product of the SiF$_2$/BF$_3$ reaction on a macroscopic scale.

Two re-examinations of the infrared spectra of matrix-isolated SiF$_2$ have been reported very recently [67,68]. This work was characterized by improved matrix isolation and by the use of both neon and argon matrices. Hastie, Hauge, and Margrave [67] established the stretching fundamentals in a Ne matrix to lie at 851 cm^{-1} (ν_1) and 865 cm^{-1} (ν_3), representing a red shift of approximately 8 cm^{-1} from the gas phase. A bond angle of 97-102° was calculated from observed isotopic splitting; molecular geometry is thus not greatly perturbed by the matrix environment. Milligan and Jacox [68], who generated SiF$_2$ from the vacuum photolysis of SiF$_2$H$_2$ or SiF$_2$D$_2$, were able to directly observe the bending fundamental of 343 cm^{-1} (in an Ar matrix). These authors also measured a series of

15

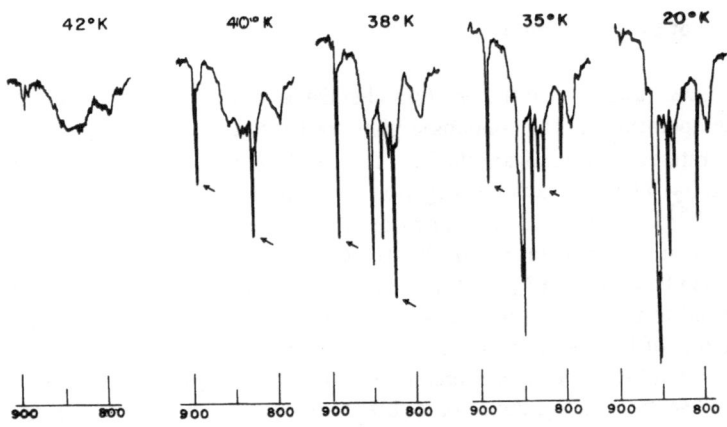

Fig. 1. Infrared spectra of SiF_2 in argon matrix during warm-up. The bands attributed to Si_2F_4 are shown by arrows

bands in the ultraviolet which correspond closely to those seen in the gas phase spectrum of SiF_2 (see Ref. [61]).

ESR Studies. Both the low-temperature chemistry and the colored appearance of SiF_2 condensate strongly suggest the presence of radical species containing unpaired electrons. Consequently, an attempt was made in this laboratory [69] to detect an electron spin resonance signal from the condensate. A gaseous SiF_2/SiF_4 mixture was condensed on a liquid nitrogen-cooled cold finger in the spectrometer cavity. The condensate generated in this manner gave rise to a broad signal whose intensity was invariant with time as long as the low temperature was maintained. The g factor for the resonance was 2.003 ± 0.002, essentially that of a free electron. When the condensate was allowed to warm, the signal decayed rapidly, and could not be regenerated by subsequent cooling—— indicating that polymerization is irreversible and complete. The nature of the signal is similar to that obtained irradiation of polytetrafluoroethylene [70].

3. Reactions of SiF_2

One of the first reactions of SiF_2 to be investigated was that with boron trifluoride [71]. The apparatus used to study this reaction is shown in Fig. 2; this appa-

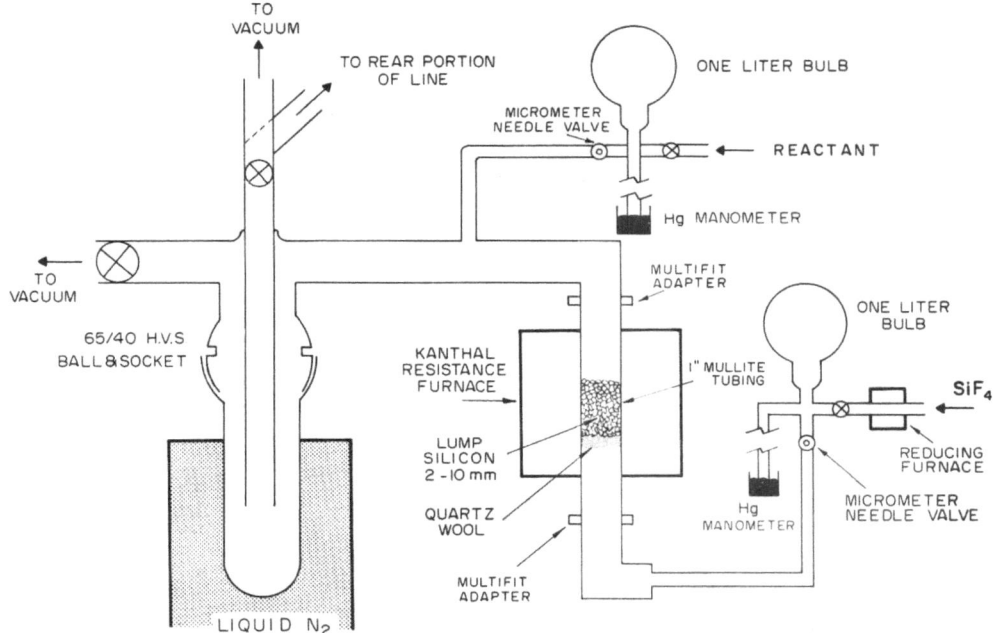

Fig. 2. Apparatus for SiF$_2$ studies

ratus is typical of those used in all of the following reactions of SiF$_2$. When SiF$_2$ and BF$_3$ are co-condensed at -196°, a green solid results. Warmup of the condensate leads to a number of volatile species, including the new compounds SiF$_3$SiF$_2$BF$_2$ and SiF$_3$(SiF$_2$)$_2$BF$_2$. The reaction products each contain at least two silicon atoms, and a gas phase reaction does not occur. These observations lead to the suggestion of the mechanism shown below:

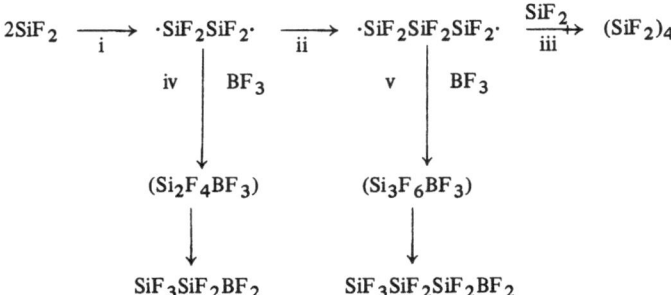

where reactions ii and iii are fast compared to iv and v. In a more recent study [72] a mixture of SiF$_2$ and SiF$_4$ was reacted with a mixture of BF and BF$_3$. The com-

17

pound $F_2Si(BF_2)_2$ was isolated from the resultant products. Diboron tetrafluoride was also reacted with SiF_2 but any new products that were formed were too unstable to be recovered.

The next series of reactions examined involved simple unsaturated and aromatic hydrocarbons and their fluorocarbon analogs. The SiF_2/benzene reaction [73] produces a series of compounds of formulae $C_6H_6(SiF_2)_n$, with $n = 2$ to at least 8. Both infrared and ultraviolet spectra indicate the absence of conjugated π systems for the $n = 3$ (highest yield) product. Hydrolysis of the product mixture gives 1,4-cyclohexadiene. These facts, along with the proton nmr spectrum of the compound, permit the conclusion that the products possess the bridged structure shown

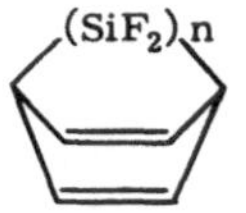

where $n = 2-8$

The reaction of SiF_2 with ethylene [74] yields the two cyclic molecules

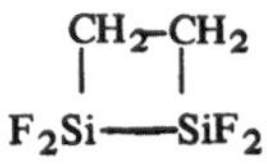

and

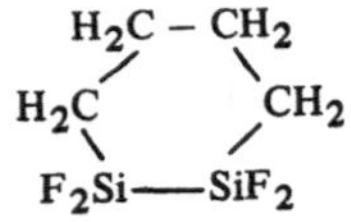

both of which are of quite limited stability. The SiF_2/acetylene [75] reaction proceeds similarly to give

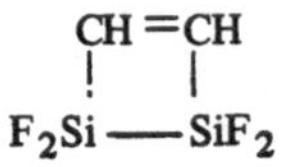

and

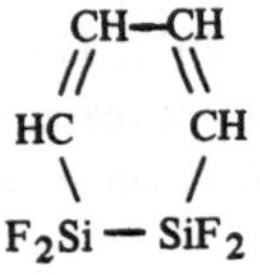

but here the six-membered ring is not isolated and is instead recovered as the rearrangement product, $HC\equiv C\text{-}SiF_2\text{-}SiF_2\text{-}CH=CH_2$. Another cyclic compound was isolated from the reaction of SiF_2 with butadiene [76].

The above reactions reinforce the "diradical" mechanism proposed for the BF_3 reaction. Hexafluorobenzene and the various fluorinated ethylenes [73,74], however, react quite differently. The products in these reactions formally correspond to C-F bond insertion by an SiF_2 *monomer*.

$$SiF_2 + \underset{F}{\overset{F}{\underset{F}{\bigcirc}}}{\overset{F}{\underset{F}{}}}F \longrightarrow \underset{F}{\overset{F}{\underset{F}{\bigcirc}}}{\overset{F}{\underset{F}{}}}SiF_3 + \underline{o}, \ \underline{m}, \ \underline{p} \ C_6F_4(SiF_3)_2$$

$$SiF_2 + CFH{=}CF_2 \longrightarrow CFH{=}CFSiF_3 + \underset{F_3Si}{\overset{H}{\underset{}{\diagup}}}C{=}CF_2$$
$$\qquad\qquad\qquad\qquad\qquad (cis \text{ and } trans)$$

Attack of a C-F bond was shown to be preferential to attack of a C=C bond.

Several quite recent investigations into SiF_2 chemistry conducted in this laboratory have further indicated the versatility of *SiF$_2$ as a reactant.* Hydrogen sulfide reacts with SiF_2 to form predominantly SiF_2HSH and Si_2F_5H [77]. The disilanethiol, SiF_2HSiF_2SH, expected from addition of H_2S to Si_2F_4, was obtained in limited yield and was observed to be quite unstable. The H_2S reaction closely paralleled an earlier study of the SiF_2/GeH_4 reaction [78] in which the products were the germylsilanes $GeH_3(SiF_2)_nH$, $n = 1-3$. The $n = 1$ homolog was the major product and compound stability decreased dramatically with increasing n.

The reaction of SiF_2 and iodotrifluoromethane was studied [79] in expectation of obtaining the products corresponding to "addition" of CF_3I to $(SiF_2)_n$ species, $CF_3(SiF_2)_nI$. The reaction of CF_3I with tetrafluoroethylene has been shown to yield (mainly) $CF_3CF_2CF_2I$ [80]. In fact, three separate homologous series of products were characterized:

$$CF_3I + SiF_2 = CF_3(SiF_2)_nI \qquad n = 1, 2$$

$$SiF_3(SiF_2)_nI \qquad n = 0, 1, 2$$

$$SiF_2I_2; \ SiF_2ISiF_2I$$

Excesses of CF_3I in the condensing mixture afforded large yields of CF_3SiF_2I; excesses of SiF_2 resulted in the formation of most or all of the compounds listed above. The CF_3I reaction is of interest as regards reaction mechanisms in SiF_2 chemistry. As in many other reactions not involving unsaturated reactants, the product obtained in highest yield contained a single silicon atom. Moreover, while each product which contained a -CF$_3$ moiety also included an I atom, the converse was not true. Similar behavior was exhibited in the H_2S reaction: SiF_3H and Si_2F_5H were products; SiF_3SH and Si_2F_5SH were not. One may make two suggestions from these observations. The first is that diradical species play a major role in SiF_2 chemistry only when there is no bond of sufficient lability to be attacked by SiF_2 monomers. Such reactive bonds include C-F in C_6F_6; C-I in CF_3I; and the O-H and S-H bonds in H_2O and H_2S.) Secondly, attack by SiF_2 or $(SiF_2)_n$ often appears to be stepwise with, in the case of CF_3I, abstraction of an I atom followed by attack on the resultant CF_3 fragment or abstraction of iodine or fluorine from another molecule.

The low-temperature condensate of SiF_2 and elemental iodine produces only SiF_2I_2 and SiF_3I, in the approximate ratio of 3:1, on warming [137]. One must consider the question of whether compounds containing silicon-silicon bonds, such as SiF_3SiF_2I or SiF_2ISiF_2I, are formed in the reaction, and suffer Si-Si bond cleavage by unreacted I_2. Although some disilanes do undergo fission reactions with I_2, the hydrogen analogs of the compounds in question (i.e., Si_2H_5I and $Si_2H_4I_2$) react only to give further substitution, eventually yielding Si_2I_6 [81]. Moreover, Si_2F_5I and $Si_2F_4I_2$, which were formed in the CF_3I/SiF_2 reaction, exhibited moderate stability in the presence of small amounts of I_2 generally present in the product mixtures of that reaction, and no evidence of these molecules was found even when large SiF_2/I_2 ratios were employed. It seems likely, then, that only SiF_2 monomers are involved in the I_2 reaction.

Hydrolysis of a silicon-halogen bond often results in formation of oxygen-containing polymers such as silicones. However the *siloxanes* Si_2OCl_6 and $Si_3O_2Cl_8$ can be recovered from careful hydrolysis of $SiCl_4$ [82a]. The reaction of SiF_4 with excess water produces fluorosilicic acid and hydrated silica, but if SiF_4 is passed over wet magnesium sulfate, one obtains the perfluorosiloxanes Si_2OF_6 and $Si_3O_2F_8$ [82b]. The controlled hydrolysis of SiF_2 might therefore be expected to lead to any of several products. Several oxygen-containing molecules other than water have been observed to react with SiF_2 to produce homologous series of both linear and cyclic oxyfluorides [83]; alternatively, insertion of an SiF_2 monomer or telomer into an O-H bond would result in formation of the silanols $H(SiF_2)_nOH$, which would almost certainly be unstable with respect to condensation to siloxanes.

The SiF_2/H_2O reaction [84] was conducted in a manner designed to minimize the contact of reactants before condensation in the cold trap. Reactions in which the SiF_2/H_2O ratio varied from 1:1 to 7:1 were conducted, but in all cases the only products not attributable to hydrolysis of SiF_4 were 1,1',2,2'-tetrafluoro-disiloxane, $HF_2Si\,OSiF_2H$, and a voluminous white polymer. No evidence for volatile compounds containing more than two silicon atoms was obtained. The polymer was of interest inasmuch as it, unlike virtually all other SiF_2 copolymers, was not pyrophoric. Infrared analysis demonstrated the absence of Si-H bonds in the polymer. The structure of the polymer is as yet unknown, but it must differ from the Si-O-F polymers formed in various other SiF_2 reactions.

The reaction of SiF_2 with methanol [85] pursued a different course from the water reaction. Here, the reaction products were CH_3OSiF_3, SiF_3H, and $(CH_3O)_2SiF_2$. Again, the competing reaction with SiF_4 represents a complication. In a separate experiment conducted under similar conditions, SiF_4 was shown to react readily with CH_3OH to form $CH_3OSiF_3 + HF$. Further methanolysis of the product to form $(CH_3O)_2SiF_2$, however, occurred only very slowly. From these and other observations, the authors formulated the following reaction scheme:

$$SiF_4 + CH_3OH = SiF_3OCH_3 + HF$$

$$SiF_2 + HF = SiF_3H$$

$$SiF_2 + CH_3OH = SiF_2HOCH_3$$

$$SiF_2HOCH_3 + CH_3OH = SiF_2(OCH_3)_2 + H_2$$

Since the Si-H bond has been observed to react readily with methanol [86], failure to observe SiF_2HOCH_3 is not surprising.

Reactions of *SiF$_2$ with NaF and LiF* have been studied [87]. The alkali fluorides were vaporized from a Knudsen cell and co-condensed with approximately equal amounts of SiF_2 on a liquid-nitrogen cooled cold finger. The reactions are complicated by gas-phase reactions and also by the reactions of SiF_4 with the fluorides. MF reactions lead to deposition of the gas phase SiF_2. The low-temperature condensate is a reddish-brown at -196°, and in all cases decomposes suddenly on warmup to deposit M_2SiF_6 and elemental silicon on the walls of the apparatus. The nature of the low-temperature solid is as yet undetermined.

Conclusions. The utility and versatility of silicon difluoride as a chemical reagent has clearly been demonstrated. Although the region of reactions of SiF_2 with small (i.e., volatile) organic molecules has been rather well covered, there remain a great number of potentially rewarding reactions with inorganic substances. Rather than dwell on this point, however, the authors would prefer to mention two potential areas for expanding SiF_2 chemistry.

No direct evidence for the observation of triplet gaseous SiF_2 exists. If such species could be generated in reasonably high yield (by mercury-sensitized photolysis, for example), their chemistry from both synthetic and kinetic points of view would merit considerable interest. Heicklen and co-workers [46] have successfully conducted similar studies with triplet CF_2 as generated from the reaction of C_2F_4 with ground-state oxygen atoms.

An aspect of Pease's early work with SiF_2 systems may have significance for the future SiF_2 chemistry. In a variation of the usual "matrix-trapping" technique, Pease studied the reaction of SiF_2 with Br_2 by passing the gases through an 8″ length of tubing heated to 1200° prior to condensation. Although the same product, SiF_2Br_2 was collected on warmup of the condensate, none of the usual room-temperature polymer was retained in the reaction trap-indicative of a quantitative gas-phase reaction. The heretofore unknown gaseous chemistry of SiF_2 might well be discovered via a similar approach on a general basis.

D. Other Silicon Dihalides

$SiCl_2$ has long been postulated to be an intermediate in gas-phase pyrolyses of various chlorosilanes [88], and in such reactions as that of Si and HCl to form $SiCl_3H$ [89], or the reduction of perchlorosilanes with H_2 [90]. Direct observation of $SiCl_2$ monomer, or successful attempts to investigate the reaction chemistry of the monomer have, however, been very sparse until quite recently. This situation is in large part due to the fact that the gas-phase lifetimes of the heavier dihalides are several orders of magnitude less than that of SiF_2. Thus, although equilibrium measurements of the system $Si + SiCl_4 = 2SiCl_2$ [91] indicate that $K_p \sim 1$ at 1350 ^oC, techniques similar to those utilized for production of SiF_2 lead to $(SiCl_2)_n$ and perchlorosilanes [92].

Timms [93] has recently studied $SiCl_2$ reaction chemistry by employing fast pumping speeds and low (5×10^{-6} torr) permanent gas pressures. Under these conditions the $SiCl_2$, which is produced from reduction of the tetrahalide with the metal at 1350 ^oC can successfully be condensed on cooled surfaces. Condensation of the equilibrium $SiCl_2/SiCl_4$ mixture at liquid nitrogen temperatures gives rise to a brown solid which turns white and evolves perchlorosilanes on warming. Co-condensation of PCl_3, BCl_3, or CCl_4 yields products corresponding to insertion of $SiCl_2$ into a M-Cl bond; products containing more than one silicon are not found. $SiCl_2$ may behave more similarly to SiF_2 (that is, diradicals may be important) in its reactions with unsaturated and aromatic compounds; Timms reported that such reactions lead to involatile polymers which incorporate the organic molecule.

Spectroscopic observations of $SiCl_2$ have been reported by Asandi, Karim and Samuel [94] and Milligan and Jacox [95]. The early work of Asandi et al. was concerned with the emission spectra from the products of an electric discharge through $SiCl_4$. The spectrum attributable to $SiCl_2$ consisted of a number of features superimposed on a continous band from 3160 to 3550 Å. The spectral features were used to tentatively assign two of the ground state vibrational fundamentals as 250 and 540 cm^{-1}. Milligan and Jacox generated $SiCl_2$ from the vacuum photolysis of SiH_2Cl_2 (or SiD_2Cl_2) in argon matrices at 14 ^oK. Examination of infrared spectra taken subsequent to photolysis revealed the stretching fundamentals to occur at 502 and 513 cm^{-1}. It was not possible to assign the symmetries of the two absorptions.

Much of the known high-temperature chemistry of silicon-chlorine compounds is indirectly concerned with $SiCl_2$. The reactions discussed below will serve as examples of those in which the intermediacy of $SiCl_2$ is indicated.

The "direct synthesis" of $SiCl_3H$ from Si and HCl has been shown [89] by kinetic studies to proceed via

$$Si + 2 HCl = SiCl_2 + H_2$$

$$SiCl_2 + HCl = SiHCl_3$$

Synthesis of organosilanes from silicon and RCl is facilitated if a silicon-copper alloy is employed [96]. The catalytic action of the copper is due to the formation of CuCl, which then reacts with silicon to form $SiCl_2$:

$$CH_3Cl + Cu \longrightarrow CH_3Cl\text{-}Cu$$

$$2\ CH_3Cl\text{-}Cu \longrightarrow 2\ CuCl + CH_4 + C + H_2$$

$$Si + 2\ CuCl \longrightarrow SiCl_2 + 2\ Cu$$

$$SiCl_2 + CH_3Cl\text{-}Cu \longrightarrow CH_3SiCl + CuCl$$

$$CH_3SiCl_2 + CH_3Cl\text{-}Cu \longrightarrow (CH_3)_2SiCl_2 + CuCl$$

The existence of $SiCl_2$ as an intermediate was indicated from an experiment in which the volatile product of a Si/CuCl reaction was allowed to react with CH_3Cl to form methylchlorosilanes.

The high-temperature (1000-2000 °K) reactions

$$SiCl_4 + 2\ H_2 = Si + 4\ HCl$$

$$SiCl_3H + H_2 = Si + 3\ HCl$$

were examined by Sirtl and Reuschel [90]. Considerations of silicon yield as a function of temperature and mole fraction of reactants lead to the conclusion that $SiCl_2$ is an important reaction intermediate.

The $SiCl_4$/Si reaction may lead to several different products, depending on reaction conditions. Thus, under conditions of high vacuum and fast pumping, $SiCl_2$ may be isolated by rapidly quenching the reaction products. Under less stringent vacuum conditions, $(SiCl_2)_n$ is deposited just beyond the hot zone, and the perchlorosilanes Si_nCl_{2n+2} can be trapped further downstream [92]. If, however, $SiCl_4$ is recycled over hot silicon in a closed system [97], viscous subchlorides of formulae Si_nCl_{2n} are obtained. The value of n varies from 12 at 900° to 16 at 1200°. The presumably cyclic compounds were characterized only by standard quantitative analyses; no spectroscopic or other physical data were reported. Bromination of the compounds produced some $SiCl_3Br$, indicating that at least some open-chain compounds were present. Since $SiCl_2$ reacts readily with $SiCl_4$ under other conditions, it is difficult to explain the apparently quantitative production of $SiCl_2$ necessary to form high-molecular weight rings. At any rate, the closed system reaction certainly merits further investigation.

Recent *thermochemical data* for $SiCl_2$ have been reported by Schäfer and co-workers [91a], and by Teichmann and Wolf [91b] from transpiration studies of the $SiCl_4$/Si system. The heat of formation of the gaseous monomer now seems well established:

$$\Delta H_f^0 (\text{SiCl}_2, \text{g}, 298 \ ^0\text{K}) \ = \ -38.2 \pm 1.5 \ \text{Kcal.}$$

Mass spectrometric investigations of the silicon-chlorine system do not seem to have been made.

Although for the dibromide and diiodide, the respective $\text{SiX}_4 + \text{Si} = 2\text{SiX}_2$ equilibria and the reaction chemistry of the $(\text{SiX}_2)_n$ polymers have been well characterized, the physical and chemical properties of the monomeric dihalides remain virtually unknown. In a preliminary report, Timms [98] related the formation of SiBr_2 in 90% yield with the apparatus and procedures employed in SiCl_2 production. The only reaction reported was that with BF_3. The sole product was BF_2SiF_3 — presumably formed from disproportionation of the expected BF_2–SiBr_2F.

Production of SiI_2 under similar (i.e., low-pressure, high-temperature) conditions is difficult due to the appreciable decomposition of SiI_2 to Si and I atoms at the temperatures required. Indeed, the diiodide has been utilized for the transportation and deposition of silicon [99].

E. Germanium Difluoride

Germanium difluoride differs dramatically from CF_2 and SiF_2 in that it can be isolated as a stable compound at room temperature. Thus, it is surprising that few of its properties have been described.

GeF_2 can be prepared by reducing GeF_4 with Ge at temperatures above 120 °C [100]. It can also be prepared by heating germanium powder with anhydrous HF (225 °C, 16 hrs) [101].

$$\text{Ge} + 2\text{HF} \longrightarrow \text{GeF}_2 + \text{H}_2$$

A number of the physical properties of GeF_2 have been measured including its infrared, ultraviolet and mass spectra. The crystal structure of GeF_2 has also been determined.

The *ultraviolet absorption spectrum* of GeF_2 has been measured by Hauge, Khanna and Margrave [102]. The spectrum is fairly simple and is probably due to the perpendicular $^1\text{B}_1 \leftarrow \text{X}^1\text{A}_1$ transition. All progressions were explained in terms of bending frequencies of the lower and upper electronic states, which are $v_2'' = 263 \ \text{cm}^{-1}$ and $v_2' = 164 \ \text{cm}^{-1}$. The 0,0,0 – 0,0,0 transition is reported to lie at 2280.1 Å.

The *infrared spectrum* of GeF_2 has also been reported [103]. It was necessary to study the matrix-isolated spectrum for two reasons. First, the examination of the ultraviolet absorption spectrum of GeF_2 indicated that at least ten of the bending states were populated, and second, germanium has five abundant isotopes. These suggested that the gas phase spectrum would be broad and ill defined at the temperatures required to vaporize GeF_2 (150 ºC). As anticipated the authors found that the gas phase spectrum of GeF_2 did consist of broad absorbances centered at 663 and 676 cm^{-1}.

The spectrum of GeF_2 trapped in a neon matrix is shown in Fig. 3. The ratio of GeF_2/rare gas in the matrix was 1:1000. When new matrices were prepared similar spectra were obtained, even when the ratio of diluent was changed or the temperature of deposition was altered. This indicated that the splitting seen in the spectrum was due to isotope effects and was not due to matrix effects. As can be seen the intensities of the various peaks are in the same ratio as the abundant isotopes of germanium, providing additional evidence that the splitting is due to isotope effects.

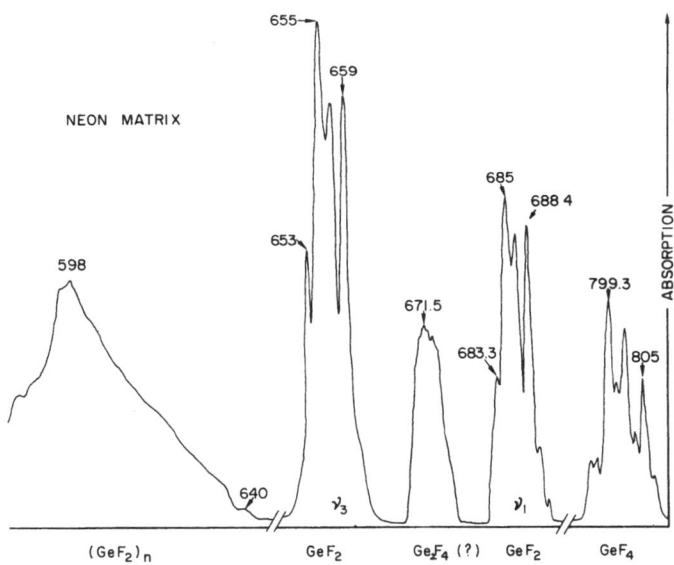

Fig. 3. IR absorption spectrum of GeF_2 matrix isolated in neon at $\sim$5 ºK

One may calculate the bond angle of GeF_2 from the isotope splitting. If one assumes that the lower frequency absorption is ν_3 the bond angle is 94 ± 4 º. If the higher frequency is ν_3 the bond angle is 82 ± 3 º. Since CF_2 and SiF_2 have bond angles of 104.9 º and 100.9 º, respectively, the value of 94 ± 4 º seems more likely to be the correct value and this has recently been confirmed by microwave studies [137].

There have been several *mass spectrometric examinations* of GeF_2 The first was by Ehlert and Margrave [65a]. GeF_2 was prepared by heating Ge and CaF_2. The ionization potential of GeF_2 was determined to be 11.6 ± 0.3 eV and the heat of atomization of $GeF_2(g)$ to be $\Delta H^0_{a,298} \sim 16.0 \pm 0.8$ eV, thence ΔH^0_f $(GeF_2, c, 298) \sim -140 \pm 17$ kcal mole^{-1}. The second investigation [104a] has required repetition since it is now known that the sample used was not pure GeF_2 [106]. In this more recent examination [37] the vapor species over pure GeF_2 were monitored over the temperature range 70–95 ^{0}C. Only GeF_2 and $(GeF_2)_2$ were found; no GeF_4 was detected. The thermodynamic data determined from this experiment are listed in Table 2.

Table 2. *Thermodynamic data from mass spectrometric experiments*

Reaction	ΔH^0_{365}	ΔS^0_{365}
$GeF_2(c) = GeF_2(g)$	19.4 ± 1.0	44.7 ± 2.5
$2 GeF_2(c) = (GeF_2)_2(g)$	18.3 ± 2.4	36.7 ± 6.0

The heat of formation of solid GeF_2 has very recently been determined by fluorine bomb calorimetry [105]. The value of $\Delta H^0_f(GeF_2, c, 298.15\ ^0K) = -157.3 \pm 1.0$ kcal mole^{-1} was determined. This is probably the best value currently available.

The crystal structure of GeF_2 was reported by Trotter, Akhtor and Bartlett [107]. They describe GeF_2 as "a strong fluorine-bridged chain polymer, in which the parallel chains are cross-linked by weak fluorine bridges. The structural unit of the strongly bridged chains is a trigonal pyramid of three fluorine atoms and an apical germanium atom." They found that the Ge-F distances are 1.79, 1.91 and 2.09 Å and the F-Ge-F angles are 85 ^{0}, 85.6 ^{0} and 91.6 ^{0}. The two fluorine atoms at 2.09 Å are equivalent and join the germanium atoms into chains. The F atoms at 1.79 Å are weakly bonded to germanium atoms in neighboring chains whose distance is 2.57 Å. The poor packing of fluorine atoms in this arrangement is due to steric activity of the non-bonding valence electron pair on the germanium. The GeF_4 group is a distorted trigonal bipyramid with four fluorine atoms and a lone pair (in the equatorial plane) around a germanium atom.

Only a few reactions of GeF_2 have been reported; however, those currently known indicate that GeF_2 is a strong reducing reagent [100b].

$$GeF_2 + I_2 \longrightarrow GeF_2I_2 \xrightarrow{\text{decomposes}} GeI_4 + GeF_4$$

$$GeF_2 + Cl_2 \longrightarrow GeF_2Cl_2 \xrightarrow{\text{decomposes}} GeCl_4 + GeF_4$$

$$2GeF_2 + SeF_4 \longrightarrow 2GeF_4 + Se$$

$$GeF_2 + SO_3 \longrightarrow \text{explosion (products are probably } GeOF_2, SO_2)$$

$$GeF_2 + H_2O \longrightarrow \text{``Ge(OH)}_2\text{''}$$

Muetterties has described some of the reactions of GeF_2 in solution [108]. He isolated the salts, $KGeF_3$ and $CsGeF_3$, by dissolving GeF_2 in concentrated solutions of KF and CsF. In solution there must be rapid exchange between F^- and GeF_3^- since the ^{19}F nmr signal from a solution containing both species is midway between the signal due to either species alone. If a solution of GeF_2 is acidified, hydrogen is released. When GeF_2 is dissolved in dimethyl sulphoxide the complex $GeF_2 \cdot OS(CH_3)_2$ is formed. No report of bond insertion or additions to multiple bonds by GeF_2 exist.

F. Other Germanium Dihalides

The other germanium dihalides have been known for a very long time. The first reported preparation of $GeCl_2$ was later withdrawn; Winkler claimed to have formed $GeCl_2$ by reacting heated germanium with HCl but the product was actually $HGeCl_3$ [109]. Moulton and Miller [110] have shown that $HGeCl_3$ is very unstable, and decomposes to $GeCl_2$ and HCl when distilled at low pressure. $GeCl_2$ can be prepared by passing $GeCl_4$ over Ge at 350 °C [111]. It is formed also by the action of AgCl on Ge [112] and by the action of Cl_2 on Ge at 650 °C [113]. When $GeCl_4$ is reduced by hydrogen, germanium subchlorides of limiting composition $GeCl_{0.9}$ are formed. When these subchlorides are distilled under vacuum at 210 °, $GeCl_2$ can be isolated [114].

GeBr_2 and GeI_2 are much easier to prepare than the difluoride or dichloride. $GeBr_2$ can be prepared by reducing $HGeBr_3$ with Zn, or by the vacuum distillation of $HGeBr_3$ [115]. GeI_2 can be very easily prepared by precipitation from Ge^{2+} solutions [116]. It can also be prepared by the action of HI on GeS [117].

Although GeF_2 has been examined by a variety of spectroscopic techniques the other dihalides have not been examined in such detail.

The *chemiluminescent emission spectrum* of $GeCl_2$ was obtained by burning $GeCl_4$ in potassium vapor using a diffusion flame technique [118]. The spectrum consisted of a series of closely spaced diffuse bands in the region 4900–4100 Å with an underlying continuum. The bands resemble those of $SnCl_2$.

These results were taken to indicate that $GeCl_2$ is non-linear in the gas phase. If the diffuse nature of the bands is due to predissociation, then the dissociation energy of the ClGe–Cl bond is less than 64 kcal.

Both the absorption and the emission ultraviolet spectra of $GeCl_2$ were observed by Hastie, Hauge and Margrave [119]. $GeCl_2$ was produced either by vaporization from liquid $GeCl_2$ or by the reduction of $GeCl_4$ with Ge. Absorption occurred between 3301–3140 Å. The transition is probably $X^1A_1 \rightarrow {}^1B_1$, as is observed for CF_2, SiF_2 and GeF_2. Bands in the spectrum were interpreted in terms of progressions in the bending frequencies of the lower and upper states. The ground state bending frequency is 162 cm^{-1} and that of the upper state is 95 cm^{-1}.

A *microwave discharge* through $GeCl_4$ vapor at low pressure produced a continuous emission from 3125 to 3341 Å, the same range as that observed in the absorption spectrum of $GeCl_2$.

The thermodynamics of the reactions

$$Ge\ (s) + GeX_4\ (g) \longrightarrow 2GeX_2\ (g)$$

$$X = Cl, Br, I$$

have been studied by a number of workers using weight loss methods, static vapor pressure measurements and mass spectrometric techniques. The mass spectrometric investigation showed that $GeCl_2$ and $GeBr_2$ do not form polymers in the gas phase in contrast with the behavior of GeF_2. The numerical results of these various investigations are summarized in Table 3 [120].

Table 3. *Heats of formation and atomization of gaseous germanium dihalides, and stabilities of Ge-X bonds, kcal mol*[-1]

Molecule	$\Delta H^0_{f,\ 298}$	ΔH^0_{atoms}	$\bar{E}(Ge-X)$
GeF_2	-136.9 ±2 [104b, 105]	266.3 ±2 [104a, 105]	133.2 ±1 [104a,105]
$GeCl_2$	- 42 ±1 [120, 121]	188 ±5 [120]	94 ±2 [120]
$GeBr_2$	- 13 ±1 [120]	164 ±5 [120]	82 ±2 [120]
GeI_2	13 ±2 [123]	142 ±5 [123]	71 ±2 [123]

Reactions of Germanium Dihalides. Due to its ease of preparation, GeI_2 prosesses the best characterized reaction chemistry of the dihalides of germanium. $GeCl_2$, $GeBr_2$ and GeI_2 all undergo the following reactions:

$$GeX_2 + X_2 \rightarrow GeX_4$$

$$GeX_2 + HX \rightleftharpoons GeHX_3$$

$GeCl_2$ and $GeBr_2$ hydrolyze to give "$Ge(OH)_2$". $GeCl_2$ begins to decompose at 75 °C, and reacts with H_2S to give GeS [124].

Just as CX_2 is formed in solution by the basic hydrolysis of CHX_3, so can GeX_2 be formed from $HGeX_3$ in solution. For example, when $HGeCl_3$ is dissolved in ether it forms the complex $2(Et_2O)HGeCl_3$ which is thought to have the ionic structure

$$[Et_2O \longrightarrow H^+ \longleftarrow OEt_2\ ]\ [GeCl_3^-]$$

and thus it readily forms $GeCl_2$. Typical reactions of the complex are

$$2(Et_2O) \cdot HGeCl_3 \longrightarrow [GeCl_2] + 2Et_2O \cdot HCl$$

$GeCl_2 + HC\equiv CH \longrightarrow \begin{array}{c} H \quad\quad H \\ \backslash \quad\quad / \\ C = C \\ \backslash\quad/ \\ GeCl_2 \end{array} \xrightarrow{\text{2 moles}} \begin{array}{l} GeXCl_2\text{-}CH=CH\text{-}GeCl_3 \\ + [\text{-}CH\text{=}CH\text{-}GeCl_2\text{-}]_n \end{array} \quad X = H, Cl$

$GeCl_2 + H_2C=CH_2 \longrightarrow \begin{array}{c} H_2C\text{-}CH_2 \\ | \quad / \\ GeCl_2 \end{array} \xrightarrow{\text{2 moles}} \begin{array}{l} GeXCl_2\text{-}CH_2\text{-}CH_2\text{-}GeCl_3 \\ + [\text{-}CH_2\text{-}CH_2\text{-}GeCl_2\text{-}]_n \end{array}$

$GeCl_2 + H_2C=CH\text{-}CH=CH_2 \longrightarrow \begin{array}{c} H_2C\text{-}CH\text{-}CH=CH_2 \\ | \quad / \\ GeCl_2 \end{array} \longrightarrow \begin{array}{c} \overset{\frown}{GeCl_2} \end{array}$

$GeBr_2$ is observed to undergo similar reactions [124]. No direct reaction was observed between carbonyl compounds and germanium dihalides [125].

The reaction of GeI_2 have received more attention. GeI_2 will react with a carbon halogen bond [126].

$$GeI_2 + RI \longrightarrow RGeI_3$$

$$R = Bu, Ph, CH_2I, I(CH_2)_2.$$

$$GeI_2 + BuBr \longrightarrow \text{trihalogenated products} \xrightarrow{EtMgBr} Et_3GeBu$$

$$GeI_2 + ICH_2OMe \longrightarrow I_3Ge\text{-}CH_2\text{-}OMe \xrightarrow{BuMgBr} Bu_3GeCH_2OMe$$

$$GeI_2 + EtCOOCH_2I \longrightarrow I_3Ge\text{-}CH_2\text{-}CO_2Et.$$

Most of these reactions were carried out in sealed tubes. GeI_2 also reacts with multiple bonds [127].

$$R\text{-}C\equiv C\text{-}R + GeI_2 \longrightarrow \left[\begin{array}{c} R \\ C \\ || \quad GeI_2 \\ C \\ R \end{array} \right]$$

$$\begin{array}{c} GeI_2 \\ R\text{-}C \qquad C\text{-}R \\ || \qquad\quad || \\ R\text{-}C \qquad C\text{-}R \\ GeI_2 \end{array}$$

$$\left[\begin{array}{c} \text{-}C = C\text{-}GeI_2\text{-} \\ | \quad\quad | \\ R \quad\quad R \end{array} \right]_n$$

$$R = H, C_6H_5$$

GeI_2 inserts into metal-metal bonds [128,129],

$$[\pi\text{-}C_5H_5Fe(CO)_2]_2 + GeI_2 \longrightarrow [\pi\text{-}C_5H_5Fe(CO)_2]_2GeI_2$$

$$(CO)_4CoCo(CO)_4 + GeI_2 \longrightarrow (CO)_4Co\text{-}GeI_2\text{-}Co(CO)_4.$$

Similar reactions have been observed for $GeCl_2$ [130]. The halide atoms in these compounds are very labile and are easily changed for groups such as -Me, OCH_3, $-SC_2H_5$, -NCS, and $-OCOCH_3$ or other halides [128,130].

GeI_2 also reacts with organo-mercury compounds [131].

$$GeI_2 + HgEt_2 \longrightarrow GeEt_2 + HgI_2$$

$$GeI_2 + HgBu_2 \longrightarrow Bu_2IGe\text{-}GeIBu_2$$
$$\text{(dissolved in acetone)}$$

GeI_2 is also of considerable importance in the transport and purification of germanium.

G. Tin and Lead Dihalides

The dihalides of tin and lead are so very well known that it is unnecessary to summarize the extensive chemical knowledge of these compounds. The chemistry of divalent tin and lead has been reviewed several times recently [132]. A few points that are relevant to the material already discussed will be made.

The ultraviolet absorption spectra of gaseous SnF_2, $SnCl_2$, PbF_2 and $PbCl_2$ have all been recently reported. For SnF_2 a weak absorption with a regular banded structure was seen at around 2425 Å. The bending frequency of the ground electronic state is 180 cm^{-1} and for the excited state is 120 cm^{-1}. For PbF_2 no discrete band system was observed; a plot of the bending frequencies of the other Group IVB difluorides against the reciprocal of their internuclear separations enabled ν_2" to be estimated as 145 cm^{-1} and ν_2' as 105 cm^{-1}. For SnF_2 and PbF_2 the $0,0,0 - 0,0,0$ transitions are estimated to occur at 40,741 and 40,560 cm^{-1}, respectively [133].

In the ultraviolet $SnCl_2$ showed a continuous absorption with a maximum intensity at about 21,044 cm^{-1} (3220 Å). The absence of discrete bands is probably due to overlapping of closely spaced diffuse bands. For $PbCl_2$ three regions of continuous absorption were observed. These had maximum intensities at 3600, 3200 and below 2916 Å. The $SnCl_2$ and $PbCl_2$ spectra were interpreted as being due to $^1A_1 \rightarrow {}^1B_1$ transitions [119].

The mass spectra of the vapours over hot SnF_2 and PbF_2 were also examined recently [134]. SnF_2 undergoes some polymerization. Species found over molten SnF_2 at 616 °K were SnF_2 79.5%, Sn_2F_4 20.5%, and Sn_3F_6 0.027%. No dimers were found over PbF_2; this was probably due to the ready decomposition of these dimers into PbF_4 and Pb. The heats of dimerization for all of the Group IVB difluorides are listed in Table 4 [135]. In the same study a number of the thermodynamic functions of the fluorides of tin and lead were measured using Knudsen cell effusion techniques [134]. These values do not agree very well with previously available data. This area has recently been surveyed and interested readers should refer to this survey [136].

Table 4. *Heats of dimerization of group IVB difluorides (kcal mole^{-1})*
$(MX_2)_n(g) \rightarrow (MX_2)_{n-1}(g) + MX_2(g)$

Molecule	$n = 2$
CF_2	76.3 ± 3
SiF_2	—
GeF_2	18.3 ± 3
SnF_2	39 ± 2
PbF_2	—

Conclusions

A varied and productive chemistry is now established for most of the Group IV dihalides. By combining high temperature and low temperature techniques, one may isolate AX_2 species and observe molecular parameters as well as physical and chemical properties. The CX_2 (carbenes) and SiX_2 (silylenes) molecules have a rich chemistry and provide new and unique opportunities for organic and organo-metallic syntheses.

Acknowledgments

Research in the areas of high temperature chemistry, fluorine chemistry, optical and mass spectroscopy and thermodynamics has been supported at Rice University by the United States Atomic Energy Commission, by the U.S. Army Research Office (Durham), by the National Aeronautics and Space Administration, by the Petroleum Research Fund of the American Chemical Society and by the Robert A. Welch Foundation. Liquid helium for low temperature nock was provided through arrangements with the U.S. offices of Naval Research.

J. L. Margrave, K. G. Sharp, and P. W. Wilson

References

1) Cottrell, T.L.: The Strengths of Chemical Bonds, 2nd ed. New York: Academic Press Inc. 1958.

2) Day, M.C., Selbin, J.: Theoretical Inorganic Chemistry, p. 212–213, New York: Reinhold 1962.

3) Cotton, F.A., Wilkinson, G.: Advanced Inorganic Chemistry, 2nd ed., p. 456. New York: Wiley and Sons 1966. – Allred, A.L., Rochow, E.G.: J. Inorg. Nucl. Chem. 5, 264 (1958). – Pauling, L.: The Nature of the Chemical Bond, 3rd ed., p. 126. Cornell Univ. Press 1959. – Sanderson, R.T.: Inorganic Chemistry, p. 78. New York: Reinhold 1967.

4) Hine, J.: Divalent Carbon. New York: The Ronald Press Co. 1964. – Kirmse, W.: Carbene Chemistry. New York: Academic Press Inc. 1964. – Parham, W.E., Schweizerm, E.E.: Org. Reactions 13, 55 (1963). – Chinoporos, E.: Chem. Rev. 63, 235 (1963).

5) Mahler, W.: Inorg. Chem. 2, 230 (1963). – Carvell, R.G., Dobbie, R.C., Tyerman, J.R.: Can. J. Chem. 45, 2297 (1964).

6) Dalby, F.W.: J. Chem. Phys. 41, 2297 (1964).

7) Heicklen, J., Knight, V., Greene, S.: J. Chem. Phys. 42, 221 (1965).

8) Mitsch, R.A.: J. Heterocyclic Chem. 1, 59 (1964).

9) Johnston, T., Heicklen, J.: J. Chem. Phys. 47, 475 (1967).

10) Apparently most fluorocarbons form CF_2 on photolysis or under the influence of an electric discharge. For a summary of the compounds used and the processes involved for producing CF_2 in this way see: Simons, J.P., Yarwood, A.J.: Nature 192, 943 (1961). – Heicklen, J., Cohen, N., Saunders, D.: J. Phys. Chem. 69, 1774 (1965).

11) Simons, J.P., Yarwood, A.J.: Nature 187, 316 (1960).

12) Batey, W., Trenwith, A.B.: J. Chem. Soc. 1961, 1388.

13) Lenzi, M., Mele, A.: J. Chem. Phys. 43, 1974 (1965).

14) Clark, H.C., Willis, C.J.: J. Am. Chem. Soc. 82, 1888 (1960).

15) Birchall, J.M., Haszeldine, R.N., Roberts, D.W.: Chem. Commun. 1967, 287. – Atkinson, B., McKeagan, D.: Chem. Commun. 1966, 189.

16) Cohen, N., Heicklen, T.: Aerospace Corp. Report TDR-469 (5250-40) -2, Nov. 20, 1964.

17) Margrave, J.L., Wieland, K.: J. Chem. Phys. 21, 1552 (1953). Nelson, L.S., Kuebler, N.A.: J. Chem. Phys. 37, 47 (1962).

18) Modica, A.P., LaGraff, J.E.: J. Chem. Phys. 44, 3375 (1966).

19) – – J. Chem. Phys. 43, 3383 (1965); 44, 1585 (1966); 45, 4729 (1966).

20) Mastrangelo, S.V.R.: J. Am. Chem. Soc. 84, 1122 (1962).

21) Craig, N.C., Hu, T., Martyn, P.H.: J. Phys. Chem. 72, 2234 (1968).

22) Venkateswarlu, P.: Phys. Rev. 77, 676 (1950).

23) Laird, R.K., Andrews, E.B., Barrow, R.F.: Trans. Faraday Soc. 46, 803 (1950).

24) Mann, D.E., Thrush, B.A.: J. Chem. Phys. 33, 1732 (1960). – Bass, A.M., Mann, D.E.: J. Chem. Phys. 36, 3501 (1962). – Trush, B.A., Zwolenik, J.J.: Trans Faraday Soc. 59, 582 (1963).

25) Simons, J.P.: J. Chem. Soc. 1965, 5406.

26) Khanna, V.M., Besenbruch, G., Margrave, J.L.: J. Chem. Phys. 46, 2310 (1967).

27) Mathews, C.W.: J. Chem. Phys. 45, 1068 (1966).

28) Milligan, D.E., Mann, D.E., Jacox, M.E., Mitsch, R.A.: J. Chem. Phys. 41, 1199 (1964).

29) Herr, K.C., Pimentel, G.C.: Appl. Opt. 4, 25 (1965).

30) Powell, F.X., Lide, D.A., Jr.: J. Chem. Phys. 45, 1067 (1966).

31) Hofman, C.W.W., Livingston, R.L.: J. Chem. Phys. *21*, 565 (1953).
32) Margrave, J.L.: J. Chem. Phys. *31*, 1432 (1954).
33) Zmbov, K.F., Uy, O.M., Margrave, J.L.: J. Am. Chem. Soc. *90*, 5090 (1968).
34) Edwards, J.W., Small, P.A.: Nature *202*, 1329 (1964). – Gozzo, F., Patrick, C.R.: Nature *202*, 80 (1964).
35) Farlow, M.W.: U.S. Patent 2, 709, 142 (May 21st, 1955).
36) Thrush, B.A., Zwolenik, J.J.: Trans. Faraday Soc. *59*, 582 (1963).
37) a) Seyferth, D., Armbrech, F.M., Jr., Schneider, B.: J. Am. Chem. Soc. *91*, 1954 (1969) and references herein;
    b) Bevan, W.I., Haszeldine, R.N., Young, J.C.: Chem. and Ind. *1961*, 789.
38) Mahler, W.: J. Am. Chem. Soc. *84*, 4600 (1962); *90*, 523 (1968).
39) Roasch, M.S.: U.S. Patent 2, 424, 667 July 29, 1947.
40) Mitsch, R.A.: J. Heterocyclic Chem. *1*, 233 (1964); J. Heterocyclic Chem. *1*, 271 (1964). – Mitsch, R.A., Neuvar, E.W.: J. Phys. Chem. *70*, 546 (1966). – Mitsch, R.A., Neuvar, E.W., Ogden, P.H.: J. Heterocylic Chem. *4*, 389 (1967). – Mitsch, R.A., Robertson, J.E.: J. Heterocyclic Chem. *2*, 152 (1965).
41) Atkinson, B., McKeagan, D.: Chem. Commun. *1966*, 189.
42) Modica, A.P.: J. Chem. Phys. *46*, 3663 (1967).
43) Andreades, S.: Chem. Ind. (London) *1962*, 732.
44) Milligan, D.E., Jacox, M.E.: J. Chem. Phys. *48*, 2265 (1968).
45) Simons, J.P., Yarwood, A.J.: Nature *187*, 316 (1960).
46) Cohen, N., Heicklen, J.: J. Chem. Phys. *43*, 871 (1965).
47) Cohen, N., Heicklen, J.: J. Phys. Chem. *70*, 3082 (1966). – Heicklen, J., Knight, V.: J. Phys. Chem. *70*, 3893 (1966).
48) a) Schmeisser, M., Schröter, H.: Angew. Chem. *72*, 349 (1960);
    b) Schmeisser, M., Schröter, H., Schilder, H., Massone, J.F.: Rosskopf, Ber. *95*, 1648 (1962).
49) Fink, C.G., Bonilla, C.F.: J. Phys. Chem. *37*, 1135 (1933).
50) Blanchard, L.P., Le Goff, P.: Can. Chem. Soc. *35*, 89 (1957).
51) Milligan, D.E., Jacox, M.E.: J. Chem. Phys. *47*, 703 (1967).
52) Andrews, L.: J. Chem. Phys. *48*, 979 (1968).
53) Steudel, R.: Tetrahedron Letters *1967*, 1845.
54) Andrews, L.: J. Chem. Phys. *48*, 972 (1968).
55) Skell, P.S., Cholod, M.S.: J. Am. Chem. Soc. *91*, 6035 (1969).
56) Tyerman, W.J.R.: Chem. Commun. *1968*, 392.
57) Pease, D.C.: U. S. Patent 2, 840, 588, 24 June, 1958.
58) Timms, P.L., Kent, R.A., Ehlert, T.C., Margrave, J.L.: J. Am. Chem. Soc. *87*, 2824 (1965).
59) Johns, J.W.C., Chantry, G.W., Barrow, R.F.: Trans. Faraday Soc. *54*, 1589 (1958).
60) Rao, D.R., Venkateswarlu, P.: J. Mol. Spectry. *7*, 287 (1961).
61) Khanna, V.M., Besenbruch, G., Margrave, J.L.: J. Chem. Phys. *46*, 2310 (1967).
62) Rao, V.M., Curl, R.F., Timms, P.L., Margrave, J.L.: J. Chem. Phys. *43*, 2557 (1965).
63) Khanna, V.M., Hauge, R., Curl, R.F., Jr., Margrave, J.L.: J. Chem. Phys. *47*, 5031 (1967).
64) Rao, V.M., Curl, R.F.: J. Chem. Phys. *45*, 2032 (1966).
65) a) Ehlert, T.C., Margrave, J.L.: J. Chem. Phys. *41*, 1066 (1964);
    b) Margrave, J.L., Kana'an, A.S., Pease, D.C.: J. Phys. Chem. *66*, 1200 (1962);
    c) Kana'an, A.S., Margrave, J.L.: Inorg. Chem. *3*, 1037 (1964).
66) Bassler, J.M., Timms, P.L., Margrave, J.L.: Inorg. Chem. *5*, 729 (1966).
67) Hastie, J.W., Hauge, R.H., Margrave, J.L.: J. Am. Chem. Soc. *91*, 2536 (1969).

33

68) Milligan, D.E., Jacox, M.E.: J. Chem. Phys. *49*, 4269 (1968).
69) Thompson, J.C., Hopkins, H.P., Margrave, J.L.: J. Am. Chem. Soc. *90*, 901 (1968).
70) Schneider, E.E.: Discussions Faraday Soc. *19*, 158 (1955). − Bruk, M.A., Abkin, A.D., Khomikovski, P.M.: Dokl. Akad. Nauk. SSSR *149*, 1322 (1963).
71) Timms, P.L., Ehlert, T.C., Margrave, J.L., Brinckman, F.E., Farrar, T.C., Coyle, T.D.: J. Am. Chem. Soc. *87*, 3819 (1965). − Margrave, J.L., Timms, P.L.: U.S. Patent No. 3, 379, 512 (1968).
72) Kirk, R.W., Timms, P.L.: J. Am. Chem. Soc. *91*, 6315 (1969).
73) Timms, P.L., Stump, D.D., Kent, R.A., Margrave, J.L.: J. Am. Chem. Soc. *88*, 940 (1966). − Margrave, J.L., Timms, P.L.: U.S. Patent No. 3, 485, 862 (1969).
74) Thompson, J.C., Margrave, J.L., Timms, P.L.: Chem. Commun. *1966*, 566.
75) − − − Liu, C.S.: unpublished results.
76) − − Inorg. Chem. *9*, in press (1970).
77) Sharp, K.G., Margrave, J.L.: Inorg. Chem. *8*, 2655, (1969).
78) Solan, D., Timms, P.L.: Inorg. Chem. *7*, 2157 (1968).
79) Margrave, J.L., Sharp, K.G., Wilson, P.W.: J. Inorg. Nucl. Chem. *32*, 1817 (1970).
80) Haszeldine, R.N.: J. Chem. Soc. *1949*, 2856, also *1953*, 3761.
81) Ebsworth, E.A.V.: Volatile Silicon Compounds, p.91. New York: Pergamon Press 1963.
82) a) Goubeau, J., Warncke, R.: Z. Anorg. Chem. *259*, 109 (1949); b) Chaigneau, M.: Compt. Rend. *266*, 1053 (1968).
83) Margrave, J.L., Sharp, K.G., Wilson, P.W.: unpublished results.
84) − − − J. Am. Chem. Soc. *92*, 1530 (1970).
85) − − − Inorg. Nucl. Chem. Letters *5*, 995 (1969).
86) Miller, W.S., Peake, J.S., Nebergall, W.H.: J. Am. Chem. Soc. *79*, 5604 (1957).
87) Harbaugh, T.W., Margrave, J.L.: unpublished results (1968).
88) Wieland, K., Heise, M.: Angew. Chem. *63*, 438 (1951). − Schwarz, R., *et al.*: Chem. Ber. *80*, 444 (1947).
89) Joklik, J., Bazant, V.: Collection Czech. Chem. Commun. *29*, 603 (1964).
90) Sirtl, E., Reuschel, K.: Z. Anorg. Allgem. Chem. *332*, 113 (1964).
91) a) Schäfer, H., Bruderreck, H., Morcher, B.: Z. Anorg. Allgem. Chem. *352*, 122 (1967); b) Teichmann, R., Wolf, E.: Z. Anorg. Allgem. Chem. *347*, 145 (1966).
92) Schmeisser, M., Voss, P.: Z. Anorg. Allgem. Chem. *334*, 50 (1964).
93) Timms, P.L.: Inorg. Chem. *7*, 387 (1968).
94) Asundi, R., Karim, M., Samuel, R.: Proc. Phys. Soc. (London) *50*, 581 (1938).
95) Milligan, D.E., Jacox, M.E.: J. Chem. Phys. *49*, 1938 (1968).
96) Golubtsov, S., Andrianov, K., Turetskaya, R., Belikova, Z., Trofinova, J., Morozov, K.: Dokl. Akad. Nauk SSSR *151*, 1329 (1963).
97) Schenk, P., Bloching, H.: Z. Anorg. Allgem. Chem. *334*, 57 (1964).
98) Timms, P.L.: Abstract of 154th ACS National Meeting.
99) Schmeisser, M., Friederich, K.: Angew. Chem. *76*, 782 (1964). − Schäfer, H.: German Patent 966, 471, August 8, 1957.
100) a) Dennis, L.M., Laubengayer, A.W.: Z. Physik. Chem. *130*, 520 (1927); b) Bartlett, N., Yu, K.C.: Can. J. Chem. *39*, 80 (1961).
101) Muetterties, E.L., Castle, J.E.: J. Inorg. Nucl. Chem. *18*, 148 (1961).
102) Hauge, R., Khanna, V.M., Margrave, J.L.: J. Mol. Spectry. *27*, 143 (1968).
103) Hastie, J.W., Hauge, R., Margrave, J.L.: J. Phys. Chem. *72*, 4492 (1968).
104) Zmbov, K.F., Hastie, J.W., Hauge, R., Margrave, J.L.: Inorg. Chem. *7*, 608 (1968).

105) — Margrave, J.L., Wilson, P.W., Adams, G.P.: J. Chem. Thermodynamics 2, 741 (1970).
106) Apparently GeF_2 reacts with GeF_4 to form either a complex or a new compound. Possibly Zmbov *et al.* used this new compound instead of GeF_2 since they prepared their GeF_2 sample in an excess of GeF_4. This reaction is being investigated.
107) Trotter, J., Akhtar, M., Bartlett, N.: J. Chem. Soc. (A) *1966*, 30.
108) Muetterties, E.L.: Inorg. Chem. *1*, 342 (1962).
109) Winkler, J.: J. Prakt. Chem. *142*, 222 (1886); *144*, 188 (1887).
110) Moulton, C.W., Miller, J.G.: J. Am. Chem. Soc. *78*, 2702 (1956).
111) Dennis, L.M., Hunter, H.L.: J. Am. Chem. Soc. *51*, 1151 (1929).
112) Lieser, K.H., Elias, H., Kohlschutter, H.W.: Z. Anorg. Allgem. Chem. *313*, 199 (1961).
113) Harner, H.R., Trahin, D.S.: U.S. Patent 2,767,052.
114) Schwarz, R., Baronetzky, E.: Naturwissenschaften *39*, 256 (1952).
115) Brewer, F.M., Dennis, L.M.: J. Phys. Chem. *31*, 1526 (1927).
116) Foster, L.S.: Inorg. Synthesis *3*, 63.
117) Flood, E.A., Foster, L.S., Pietrusza, E.W.: Inorg. Syn. *2*, 106.
118) Tewarson, A., Plamer, H.B.: J. Mol. Spectry. *22*, 117 (1967).
119) Hastie, J.W., Hauge, R.H., Margrave, J.L.: J. Mol. Spectry. *29*, 152 (1969).
120) Uy, O.M., Muenow, D.W., Margrave, J.L.: Trans. Faraday Soc. *65*, 1296 (1969) and references therein.
121) Sedgwick, T.O.: J. Electrochem. Soc. *112*, 496 (1965).
122) Feber, R.C.: U.S.A.E.C. Rept. LA-3164 (1965).
123) Jona, F., Lever, R.F., Wendt, H.R.: J. Elctrochem. Soc. *111*, 413 (1964). — Evans, D.F., Richards, R.E.: J. Chem. Soc. 1292 (1952).
124) Various of these reactions are reviewed by Nefedov, O.M., and Manakov, M.N.: Angew. Chem. Intern. Ed. Engl. *5*, 1027–9 (1964).
125) Kolesnikov, S.P.,Perl'mutter, S.L., Nefedov, O.M.: Dokl. Akad. Nauk SSSR *180*, 112 (1968).
126) Lesbre, M., Mazerolles, P., Manuel, G.: Compt. Rend. *257*, 2303 (1963).
127) Volpin, M.E., Koreshkov, Y.D., Dulova, V.G., Kursanov, D.N.: Tetrahedron *18*, 107 (1962). — Johnson, F., Gohlke, R.S., Nasutavicus, W.A.: J. Organometal. Chem. ., 233 (1965).
128) Flitcroft, N., Harbourne, D.A., Paul, I., Tucker, P.M., Stone, F.G.A.: J. Chem. Soc. (A) *1966*, 1130.
129) Patmore, D.J., Graham, W.R.G.: Inorg. Chem. *5*, 1405 (1966).
130) Nesmeyanov, A.N., Anisimov, K.N., Kolobova, N.E., Denisov, F.S.: Izv. Akad. Nauk SSSR, Ser. Khim. *1968*, 142.
131) Jacobs, G.: Compt. Rend. *238*, 1825 (1954). — Emel'yanova, L.I., Vinogradova, V.N. Makarova, L.G., Nesmeyanov, A.N.: Izv. Akad. Nauk SSSR, Otd. Khim. Nauk *1962*, 53.
132) For a recent review see Donaldson, J.D.: Progress in Inorganic Chemistry *8*, 287– 356. New York: Interscience 1967.
133) Hauge, R.H., Hastie, J.W., Margrave, J.L.: J. Phys. Chem. *72*, 3510 (1968).
134) Zmbov, K., Hastie, J.W., Margrave, J.L.: Trans. Faraday Soc. *64*, 861 (1968).
135) – – – High Temperature Technology, p. 345. London: Butterworths 1969.
136) Adams, G.P., Margrave, J.L., Steiger, R.P., Wilson, P.W.: J. Chem. Thermodynamics *3*, 291 (1971).
137) Margrave, J.L., Sharp, K.G., Wilson, P.W.: J. Inorg. Nucl. Chem. *32*, 1813 (1970).

Received September 18, 1970

The Chemistry of Iminoboranes

Dozent Dr. A. Meller

Institut für Anorganische Chemie, Technische Hochschule Wien, Austria

Contents

I. Introduction

The chemistry of iminoborane compounds containing the $>C=N-B<$ moiety has developed only within the last decade. The first representatives of this type of compounds were obtained by hydroboration of nitriles with sterically hindered boranes [10] or tetraalkyldiboranes [17]; the resultant compounds appeared to be unique intermediates (stabilized by steric or reactivity effects) in the course of reactions that normally lead to borazines. The intermediates illustrated in Eq. (1) are mostly unstable at room temperature and, in general, cannot be isolated.

$$3\ R-C{\equiv}N + 1{,}5\ B_2H_6 \longrightarrow 3\ R-C{=}N{:}BH_3 \longrightarrow \frac{3}{n}\ [R-CH{=}N-BH_2]_n \longrightarrow$$

$$\underset{\displaystyle \overset{RCH_2-N}{}\ \ \underset{\displaystyle \overset{B}{H}}{\diagdown}\ \ \overset{\displaystyle N-CH_2R}{}}{\overset{\displaystyle \overset{CH_2R}{\underset{\textstyle |}{N}}}{HB \qquad BH}} \tag{1}$$

However, the successful isolation of an iminoborane derivative in the reaction of trichloroacetonitrile with diborane indicated that the stability of iminoboranes is not only a function of the nature of the borane but also of that of the imine [16].

It is now recognized that iminoboranes are not merely unique intermediates but rather are members of an independent class of stable compounds which has its own chemistry such as the aminoboranes, borazines, or other examples of boron-nitrogen compounds.

To date, monomeric (I) and dimeric (II) iminoboranes have been identified. These show a very different behaviour due to their difference in hybridization at the boron atoms.

$$>C=N\cdots B< \qquad\qquad \text{(II)}$$
$$\text{(I)}$$

Dimeric iminoboranes are generally solids at room temperature and many of them are quite stable towards hydrolysis due to the tetra-coordination about the boron atoms. Substituted dimeric iminoboranes can produce geometrical isomers as was established by ^{1}H or ^{19}F n.m.r. spectroscopy [4] and in one case such isomers have been separated [17]. The monomeric iminoboranes containing three-coordinate boron are much more reactive than the monomers. Nevertheless etheric solvents slowly interact with many halosubstituted dimeric iminoboranes and should not be used.

Most of the monomeric compounds are highly reactive liquids and are rapidly hydrolysed by atmospheric moisture. In the monomeric iminoboranes the N-B bond order is greater than unity and these compounds represent an allene-type system with cumulated multiple bonds as indicated by structure (I) [35].

Equilibria involving both forms [6,24,31,35] as well as equlibria between the dimeric form and the nitrile-borane adducts, (III) [24,31], are known to exist.

$$-C \equiv N : B< \qquad (III)$$

Transitions from one form to the other (e.g., from (II) $\rightleftarrows$ (I) or (II) $\rightleftarrows$ (III) due to temperature changes have been observed in several cases and will be noted in the following presentation.

II. General Survey of Synthetic Procedures

Iminoboranes can be prepared by several methods, depending on their substituents and the availability of suitable starting materials.

1. The most widely used method for the synthesis of iminoboranes involves the 1,2-addition of boron-element bonds such as boron-hydrogen, boron-halogen, boron-carbon, or boron-sulfur bonds across the C≡N bond of nitriles thereby producing variously substituted iminoboranes (Eq. (2)). The formation of iminoboranes as well as the stability of the products depends on the substituent on the nitrile group, the nature of the boron-element bond to be cleaved during the 1,2-addition across the C≡N bond, and to a lesser extent on the non-reacting boron substituents [26].

$$R-C \equiv N + XBX_2 \longrightarrow \underset{\overset{|}{X}}{R-C}=N-BX_2 \qquad (2)$$

If the boron compound contains more than one reactive bond several means exist to elucidate which bond has been split.

a) Some indication about the reaction course is provided by the reaction rate. The 1,2-addition of B—H compounds to C≡N groups appears to proceed readily but intermediate adducts have been reported [8]. The 1,2-addition of boron-halogen bonds to nitriles occurs almost instantaneously even at low temperatures, if it can be achieved [23,26,31], and no intermediate products have been observed. Those reactions involving cleavage of B–C bonds have been found to proceed more readily at elevated temperatures [5,6,40] (exception: triallylborane [2])) and corresponding adducts might exist at low temperatures [40]. Additions of B-S compounds across C≡N triple bonds are very slow reactions which proceed at room-temperature during a period of several days. An increase in the reaction temperature fails to enhance the formation of the desired compounds [32,33].

b) Infrared spectroscopy is a useful tool in many cases to determine the product structures, and n.m.r. spectra, particularly those of unsymmetrically substituted compounds, can substantiate the structures. Fragmentation in the mass spectrometer can also be helpful but should be used only in conjunction with other methods, since intramolecular group transfer seems to occur quite frequently [14,29].

2. The reaction of diphenylketimine-lithium with haloboranes yields a great number of diphenylketiminoboranes, Eq. (3).

$$(C_6H_5)_2C{=}N{-}Li + XB{<} \longrightarrow (C_6H_5)_2C{=}N{-}B{<} + LiX \qquad (3)$$

Diphenylketimine-lithium has been advantageously replaced by diphenylketiminotrimethylsilane. The resultant trimethylhalosilanes are more readily separated from the iminoboranes than are the lithium salts (Eq. 4).

$$(C_6H_5)_2C{=}N{-}Si(CH_3)_3 + XB{<} \longrightarrow (C_6H_5)_2C{=}N{-}B{<} + (CH_3)_3SiX \qquad (4)$$

A few compounds have also been prepared by reacting iminehydrochlorides with sodium tetraphenylborate [45,47] as illustrated be Eq. (5)

$$>{C}{=}NH_2Cl + NaB(C_6H_5)_4 \longrightarrow >{C}{=}N{-}B(C_6H_5)_2 + NaCl + 2\ C_6H_6 \qquad (5)$$

3. Several iminoboranes have been prepared by special procedures. For example, iminoboranes which are derivatives of hexafluoroisopropylideneimine, $(CF_3)_2CNH$, are obtained by dehydrohalogenation of the (amino)haloboranes with an excess of the isopropylideneimine [37] (Eq. 6).

$$(CF_3)_2CBr{-}NH{-}B(C_6H_5)_2 + (CF_3)_2CNH \rightarrow$$
$$(CF_3)_2C{=}N{-}B(C_6H_5) + (CF_3)_2CBr{-}NH_2 \qquad (6)$$

The compound $[Cl_2C{=}N{-}BCl_2]_2$ is formed on photochemical chlorination of 1,3,5-trimethyl-2,4,6-trichloroborazine [30] or by interaction of trichloroborane with thiocyanogen trichloride. The reaction of the latter compound with tribromoborane leads to $[Cl_2C{=}N{-}BBr_2]_2$ [21]. These processes are summarized in Eq. (7).

$$\begin{array}{c} \underset{CH_3}{\underset{|}{N}} \\ \underset{ClB}{} \overset{}{\diagdown} \underset{BCl}{} \\ CH_3N \diagdown \underset{\underset{Cl}{|}}{B} \diagup NCH_3 \end{array} \quad \xrightarrow[CCl_4, \text{ u.v.}]{Cl_2} \quad \begin{array}{c} Cl \diagdown \underset{C}{} \diagup Cl \\ \parallel \\ Cl \diagdown \underset{B}{} \diagdown \underset{N}{} \diagdown \underset{B}{} \diagup Cl \\ Cl \diagup \diagdown \underset{N}{} \diagdown \diagdown Cl \\ \parallel \\ Cl \diagup C \diagdown Cl \end{array}$$

$$\xrightarrow{BCl_3}$$

$$(7)$$

$$Cl-S-N=CCl_2 \qquad \xrightarrow{BBr_3} \qquad \begin{array}{c} Cl \diagdown \underset{C}{} \diagup Cl \\ \parallel \\ Br \diagdown \underset{B}{} \diagdown \underset{N}{} \diagdown \underset{B}{} \diagup Br \\ Br \diagup \diagdown \underset{N}{} \diagdown \diagdown Br \\ \parallel \\ Cl \diagup C \diagdown Cl \end{array}$$

$(C_6H_5)_2C=N-B(CH_3)_2$ was obtained by pyrolysis of the adduct of diphenyl-ketimine with trimethylborane [40] according to Eq. (8):

$$n\ (C_6H_5)_2C=NH:B(CH_3)_3 \quad \xrightarrow{160-200^\circ} \quad [(C_6H_5)_2C=N-B(CH_3)_2]_n + n\ CH_4$$

(for n c.f. Sec. V). $\hspace{8cm}$ (8)

III. Iminoboranes Derived from Borine or Alkylboranes

(Alkylideneamino)-butylboranes, the first known dimeric iminoboranes obtained from hydroboration reactions of nitriles [10], were prepared according to Eq. (9).

$$2\ RCN + 2\ (CH_3)_3CBH_2:N(CH_3)_3 \quad \xrightarrow[\text{diglyme}]{100^\circ} \quad \begin{array}{c} CHR \\ \parallel \\ H \diagdown \underset{B}{} \diagdown \underset{N}{} \diagdown \underset{B}{} \diagup H \\ (CH_3)_3C \diagup \diagdown \underset{N}{} \diagdown \diagup C(CH_3)_3 \\ \parallel \\ CHR \end{array} +$$

$$+ 2\ (CH_3)_3N \qquad (9)$$

$R=R=CH_3, C_2H_5, \text{n-}C_3H_7, \text{i-}C_3H_7, C_6H_5, 4\text{-}CH_3O-C_6H_4, 4\text{-}Cl-C_6H_4,$

$4\text{-}F-C_6H_4, 4\text{-}CH_3-C_6H_4, 3\text{-}CH_3-C_6H_4$

It is assumed [10] that the resultant products have a preferred sterical arrangement rather than being a mixture of possible stereoisomers. In the case of R=aryl the major products are contaminated by small amounts of higher melting materials which are supposed to be the trimeric iminoboranes.

Tetraalkyldiboranes were shown to react with acetonitrile to yield dimeric iminoboranes [17] according to Eq. (10) whereas 1,3,5-triethylborazine was obtained as a by-product when 1,1-dialkyldiborane was used in the same reaction.

$$2 CH_3CN + R_2BHBHR_2 \longrightarrow [CH_3CH=N-BR_2]_2 \qquad (10)$$

$R = R=CH_3$ or C_2H_5

A variety of substituted dimeric iminoboranes were obtained from reaction (10) as a mixture of geometrical isomers; however, only dimeric ethylideneaminodimethylborane could be separated (by vacuum distillation) into the cis- and trans-isomers (IV) and (V).

(IV) m.p. −5 °C   (V) m.p. 76 °C

Although dimeric (t-butylmethyleneamino)dibutylborane can be prepared by the reaction of tri-t-butylborane with pivalonitrile at temperature above 150 °C by elimination of butene [5], preparation at lower temperatures involving triethylamine-borane according to Eq. (11), illustrates that the formation of this iminoborane proceeds via di-n-butylborane as an intermediate [6].

$$4 [(CH_3)_3C]_3B + 2 (C_2H_5)_3N:BH_3 + 6 (CH_3)_3CCN \xrightarrow{110-130°}$$

$$3 [(CH_3)_3CCH=N-B(t-C_4H_9)_2]_2 + 2 (C_2H_5)_3N \qquad (11)$$

(t-Butylmethyleneamino)di-t-butylborane monomerizes at elevated temperatures and in vacuum.

Several arylsubstituted dimeric iminoboranes have been prepared by the reaction of (phenylmethyleneamino)trimethylsilane with diorganohaloboranes or organodihaloboranes [47] according to Eq. (12).

$$2\ C_6H_5CH=N-Si(CH_3)_3 + 2\ RBXY \longrightarrow$$

$$+ 2\ (CH_3)_3SiX \quad (12)$$

R = Y = C_6H_5, X = Cl, R = C_6H_5, X = Y = Cl, R = X = Mesit., X = F

The compound $Cl_3CH=NBH_2$ is obtained upon reaction of trichloroaceton-itrile with diborane during the synthesis of 1,3,5-trichloroethylborazine (Eq. (13) [16]); it is undoubtedly not monomeric as indicated by its i.r. spectrum ($\nu\ BH_2$ cm^{-1}, ν C=N 1705 cm^{-1} [22])) Most likely the material is dimeric and it is difficult to isolate in pure form due to the ready conversion to the bor-azine derivative.

$$3\ Cl_3CC\equiv N + 1,5\ B_2H_6 \longrightarrow \frac{3}{n}\ [Cl_3CCH=N-BH_2]_n \longrightarrow$$

(13)

Iminoboranes derived from borine or alkylboranes and related aldimino-boranes are listed in Table 3.

IV. Iminoboranes Derived from Halo- and Organohaloboranes

1. 1,2-Addition of Haloboranes Across the CN Triple Bond of Nitriles

Prior to 1958 only nitrile-adducts were reported to result from the reaction of nitriles with trihaloboranes (c.f. [36]). The first reported example of the 1,2-addition of boron-halogen bonds to a C≡N group involves the reaction of tri-fluoroacetonitrile with trichloroborane or tribromoborane [4] and leads to di-meric derivatives. A 1 : 1 ratio of the cis-trans isomers is obtained as indicated on the basis of ^{19}F n.m.r. studies.

$$2\,CF_3C{\equiv}N + 2\,BX_3 \longrightarrow \text{[structure]} \qquad X = X = Cl \text{ or } Br \qquad (14)$$

Pentafluorobenzonitrile, however, appeared to react only with the formation of a nitrile-trihaloborane adduct [4] (but c.f. Eq. (18)).

The observation that 1,3,5-trimethyl-2,4,6-trichloroborazine yields some dimeric (dichloromethyleneamino)dichloroborane upon photochemical chlorination [30] (c.f. Eq. (7)) prompted the synthesis of this compound by the reaction of BCl_3 with $ClSNCCl_2$ [21] (c.f. Eq. (7)) and also by the 1,2-addition of trichloroborane to chlorocyanide [23] as depicted in Eq. (15).

$$2\,ClC{\equiv}N + 2\,BCl_3 \longrightarrow [Cl_2C{=}N{-}BCl_2]_2 \qquad (15)$$

Reaction (15) was originally performed more than a century ago but as that time the product was formulated as the chlorocyanide-trichloroborane adduct [18]. Analogous 1,2-additions of tribromoborane to chlorocyanide and of trichloroborane to bromocyanide result in the formation of substituted dimeric iminoboranes [23].

When chloro- or bromocyanide are reacted with phenyldichloroborane a B–Cl bond [23] is added across the cyano group.

$$2\,Y{-}C{\equiv}N + 2\,C_6H_5BCl_2 \longrightarrow [Y(Cl)C{=}N{-}B(C_6H_5)Cl]_2 \qquad (16)$$

Y = Cl or Br

The same event is true for the reaction of chlorocyanide with n-buthyldichloroborane and di(n-butyl)chloroborane respectively [24] (c.f. Eq. (23) for the reaction of bromocyanide).

Studies of the 1,2-addition of trihalo-, organodihalo- and diorganohaloboranes to the nitrile groups of variously substituted nitriles have shown that boron-halogen bonds will undergo 1,2-addition if the CN groups are bonded to electron attracting substituents [26]. If the polar resonance structure (VI) is favored, 1,2-addition to the nitrile group will occur [26].

$$X{-}C{\equiv}N \longleftrightarrow \overset{\delta-}{X}-\overset{\delta+}{C}=\overset{\delta-}{N} \qquad (VI)$$

This event is clearly demonstrated by the behavior of 4-cyanopyridine which forms an adduct even with an excess of tribromoborane. However, 4-cyanopyridinium chloride yields a saltlike dimeric iminoborane derivative [26] as depicted in Eq. (17).

$$2\,Cl^- \; HN{=}{<}C_6H_4{>}{-}CN + 2\,BBr_3 \longrightarrow \left[HN{<}...{>}C{=}N\overset{Br_2}{\underset{Br_2}{B}}\overset{Br}{\underset{B}{N}}{=}C{-}{<}...{>}NH \right]^{2+} + 2\,Cl^- \qquad (17)$$

It has been found [31] that the addition of B-Cl, B-Br and B-I bonds across the nitrile group of α-halogenated alkylnitriles occurs regardless whether or not mono-, di- or tri-halosubstituted alkylnitriles are used, if F, Cl or Br are the halosubstituents on the alkylnitrile. Various monomeric and dimeric iminoboranes have been obtained by this procedure. Monomeric derivatives, sometimes in equilibrium with their dimers, result when B-Br bonds of organodibromoboranes or diorganobromoboranes are added across the CN group of trichloro- or tribromoacetonitrile [31]. For example, a monomeric compound has been obtained from the reaction of hydrogen cyanide and tribromoborane [24].

$$HCN + BBr_3 \longrightarrow \begin{array}{c} H \\ Br \end{array}\!\!\!\!>\!\!C=N-BBr_2 \qquad (18)$$

Fluorocyanide was found to produce similar monomers [26]. The iminoborane dimers obtained from chlorocyanide or bromocyanide with dimethylbromoborane transform to monomers upon vacuum distillation.

$$XCN + 2\,(CH_3)_2BBr \xrightarrow[CCl_4]{-20°} [X(Br)C=N-B(CH_3)_2]_2 \xrightarrow[11\ Torr]{120°}$$

$$2\,X(Br)C=N-B(CH_3)_3 \qquad (19)$$

The liquid monomer dimerizes to a solid upon storing at room temperature [24].

In some cases monomerization of dimeric iminoboranes (e.g., trichloromethylchloromethyleneamino)dichloroborane and (trichloromethylbromomethyleneamino)dibromoborane) can be observed on dissolving the materials in chlorinated saturated hydrocarbons. These compounds are monomeric in the gas phase and consequently have an unexpectedly low boiling point [24].

Transformations of dimeric iminoboranes into the corresponding nitrileborane adducts, due to temperature changes during sublimation or distillation, have been observed for the reaction products of pentafluorobenzonitrile and tribromoborane or organodibromoboranes [24]:

$$2\,C_6F_5CN + 2\,RBBr_2 \xrightarrow[CCl_4]{-20\ °C} \begin{array}{c} F_5C_6 \diagdown\ C \diagup Br \\ \| \\ Br \diagdown\ N \diagup R \\ \diagdown B \diagdown\ \diagup B \diagdown \\ R \diagup\ N \diagup\ Br \\ \| \\ F_5C_6 \diagup C \diagdown Br \end{array} \xrightarrow[0,001\ Torr]{100\ °C}$$

$$2\,C_6F_5C\equiv N:B(R)Br_2 \qquad (20)$$

R = R = C_6H_5, CH_3 or Br

In general, addition across the nitrile group is more favored if B-Br rather than B-Cl bonds are involved. The tendency to form iminoboranes decreases with increasing size of the organic substituent of the borane, i.e., in the sequence BX_3 (trihaloboranes) $>$ RBX_2 (organodihaloboranes) $>$ R_2BX (diorganohaloboranes) [26]. Consequently, with respect to the reactivity of the cyano compound, all stages between 1,2-addition, adduct formation and non-reaction can be realized. While chlorocyanide reacts with pentafluorophenyldichloroborane to yield the expected dimer (Eq. (21)).

$$2\ C_6F_5BCl_2 + 2\ ClCN \longrightarrow [Cl_2C=N-B(C_6F_5)Cl]_2 \tag{21}$$

analogous reactions with bromocyanide or chloromethylcyanide yield mixtures of dimeric iminoboranes and the corresponding nitrile-borane adducts at room temperature [26].

$$X = X = Br\ or\ CH_2Cl$$

A similar equilibrium between reaction products is observed when n-butyldibromoborane is reacted with bromocyanide, whereas di-n-butylbromoborane forms only an adduct with the latter [26].

A B-Cl bond of pentafluorophenyldichloroborane adds quantitatively across the $C\equiv N$ group of trichloroacetonitrile and yields an equilibrium mixture of the monomeric and dimeric iminoborane derivative [26]. In contrast, benzonitrile does not react at all with trichloroacetonitrile under comparable conditions [26].

Mixtures of dimeric iminoboranes with iodoacetonitrile-trihaloboranes were obtained from iodoacetonitrile and trichloro- or tribromoborane; triiodoborane formed only the adduct [31]. Trichloroborane has been shown to give 1,2-addition to an oxyalkylnitrile after intermediate formation of an adduct:

$$2\ CH_3OCH_2CN + 2\ BCl_3 \xrightarrow{-35\ ^\circ C} 2\ CH_3OCH_2CN:BCl_3 \xrightarrow{20\ ^\circ C}$$
$$[CH_3OCH_2(Cl)=N-BCl_2]_2 \tag{24}$$

However, the resultant product decomposes within a few days by cleavage of the ether linkage [33].

2. Reactions of Diphenylketimino Compounds

1,2-Addition of B–F bonds across C≡N groups has never been achieved. Therefore, the only B-fluoro substituted iminoborane so far known was obtained by the reaction of diphenylketimine-lithium with trifluoroborane [13] according to the general Equation (25).

$$n\ (C_6H_5)_2C=NLi\ +\ n\ BX_3\ \longrightarrow\ [(C_6H_5)_2C=N-BX_2]_n\ +\ n\ LiX \qquad (25)$$

In the case of X = Cl, Br, or I, n is 2, whereas the fluorocompound seems to be a polymer, (VII) [13].

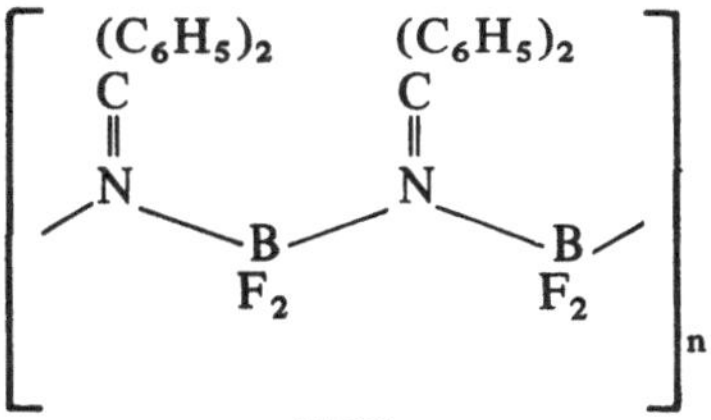

$$
\begin{bmatrix}
\begin{matrix}
(C_6H_5)_2 & & (C_6H_5)_2 \\
C & & C \\
\| & & \| \\
N & & N \\
\diagdown\!B & \diagup\!\diagdown & \!B\!\diagup \\
F_2 & & F_2
\end{matrix}
\end{bmatrix}_n
$$

(VII)

Due to difficulties in separating (VII) from the LiCl byproduct, it was found to be advantageous to prepare (VII) by the reaction of a tris(imino)borane with trifluoroborane-etherate as illustrated in Eq. (26).

$$n\ [(C_6H_5)_2C=N]_3B\ +\ n\ 2\ BF_3 \cdot O(C_2H_5)_2\ \longrightarrow$$

$$3\ [(C_6H_5)_2C=NBF_2]_n\ +\ 2n\ (C_2H_5)_2O \qquad (26)$$

Tris(diphenylmethyleneamino)borane was obtained according to Eq. (30). Exchange reactions of monomeric and dimeric iminochloroboranes with various fluorides could not be effected. However, this lack of reaction is to be expected, since it is well known from borazine chemistry that substitutions which would require a structural change in order to obtain a stable product will not occur [20]. (Diphenylketimino)trimethylsilane (rather than diphenylketimine lithium) has been utilized successfully to prepare (diphenylmethyleneamino) dihaloboranes [46]. This type of reaction is in accordance with earlier observations concerning the cleavage of Si-N bonds by haloboranes [38].

$$n\ (C_6H_5)_2C=NSi(CH_3)_3\ +\ n\ BX_3\ \longrightarrow\ [(C_6H_5)_2C=NBX_2]_n\ +\ n\ CH_3SiX$$

$$(27)$$

Trifluoroborane-etherate was used in this reaction. The same route seems to be quite convenient for the preparation of many other (diphenylmethylene-amino)boranes including dihaloborane derivatives. Similarly, B-monohalo deriv-. atives such as $[(C_6H_5)_2C=N-B(C_6H_5)Cl]$ and B-diaryl derivatives were synthesized by the cited method [47]

By utilization of more than one equivalent of (diphenylmethylenimino)trimethylsilane per haloborane several di- and triimino-substituted monomeric iminoboranes were perpared [46]. Stable bis(diphenylmethyleneamino)phenylborane was formed as depicted in Eq. (28).

$$2 (C_6H_5)_2C=N-Si(CH_3)_3 + C_6H_5BCl_2 \longrightarrow$$

$$[(C_6H_5)_2C=N]_2BC_6H_5 + 2 (CH_3)_3SiCl \tag{28}$$

The diphenylketimine lithium procedure [47] provides another route to the same compound.

Bis(imino)haloboranes, however, were found to be unstable with respect to rearrangement into (imino)dihaloboranes and tris(imino)borane; also, utilization of more than one equivalent of $(C_6H_5)C=N-Si(CH_3)_3$ converts trihaloboranes to mixtures of mono- and tris(imino)boranes which are the final products depending upon the ratio of starting materials.

$$2 n (C_6H_5)_2C=N-Si(CH_3)_3 + n BX_3 - 2 n(CH_3)_3SiX \longrightarrow \tag{29}$$

$$n[(C_6H_5)_2C=N]_2-BX \longrightarrow [(C_6H_5)_2C=NBX_2]_n + [(C_6H_5)_2C=N]_3B$$

An analogous reaction with phenyldichloroborane leads to dimeric (diphenylmethlene-amino)phenylchloroborane [47]. A similar iminoborane was refered to in Sect. III (Eq. (12)).

Pure tris(diphenylmethyleneamino)borane is readily obtained by treating tribromoborane with diphenylketimine lithium [42]:

$$3 (C_6H_5)_2C=NLi + BBr_3 \longrightarrow [(C_6H_5)_2C=N]_3B + 3 LiBr \tag{30}$$

A unique compound, (VIII), is formed be reacting $C_6H_4O_2BCl$ with $(C_6H_5)_2C=NSi(CH_3)_3$ [46] or $(C_6H_5)_2C=NLi$ [47], compound (VIII)

(VIII)

represents the only known example of a B-O substituted iminoborane.

3. Miscellaneous Reactions

Bis(trifluoromethyl)bromomethylamino-diphenylborane is readily dehydro-halogenated with an excess of hexafluoroisopropylideneimine (c.f. Eq. (6)) [37].

The bis(trifluoromethyl)methyleneaminodiphenylborane thus formed is mono-meric. Tris(ditrifluoromethylmethyleneamino)borane was found as a byproduct but was not isolated in pure state.

This cited reaction illustrates that the C=N double bond of iminoboranes is quite stable. Indeed, the C=N bond in these compounds tends to increase its bond order, forming corresponding nitriles, rather than to undergo further 1,2-additions leading to aminoboranes. This suggestion is confirmed by several reported transformations of iminoboranes to nitrile-borane adducts (Eq. (20)) [24]. Addition across the C=N double bond of iminoboranes is virtually unknown. This event is also true for related imines (e.g., dichloromethylenealkyl-amines) which yield imine-trihaloborane adducts with trihaloboranes rather than to undergo a 1,2-addition (c.f. Sect. VII).

B-halogenated iminoboranes were also obtained by treating (t-butylmethyl-eneamino)di-n-butylborane (c.f. (III) with trihaloboranes at elevated tempera-tures [6].

$$2 \text{ t-}C_4H_9(H)C=N-B(n-C_4H_9)_2 + 2 \text{ BX}_3 \xrightarrow{90 - 130 \,^\circ C}$$

$$[\text{t-}C_4H_9(H)C=N-BX_2]_2 + 2 \text{ n-}C_4H_9BX_2 \tag{31}$$

X = Cl or Br

Preparations of (dichloromethyleneamino)dichloroborane have been mentioned earlier (c.f. Eq. (7)).

4. Substitution of Halogen

Monomeric bromosubstituted iminoboranes in hydrocarbon solutions react quantitatively diphenyldiazomethane [28]. Monomeric diphenylbromomethyl-substituted iminoboranes result from this reaction as illustrated by Eq. (32).

$$\begin{array}{c} Cl_3C \\ Br \end{array}\!\!\!>\!\!C=N-B\!\!<\!\!\begin{array}{c} R \\ R \end{array} + (C_6H_5)_2CN_2 \longrightarrow \begin{array}{c} Cl_3C \\ (C_6H_5)_2C \end{array}\!\!\!>\!\!C=N-B\!\!<\!\!\begin{array}{c} R \\ R \end{array}$$
$$\qquad\qquad\qquad\qquad\qquad\qquad\qquad\qquad\qquad\qquad\qquad Br$$

R = CH_3, C_6H_5

$$\begin{array}{c} Cl_3C \\ Br \end{array}\!\!\!>\!\!C=N-B\!\!<\!\!\begin{array}{c} R \\ Br \end{array} + 2(C_6H_5)CN_2 \longrightarrow \begin{array}{c} Cl_3C \\ (C_6H_5)_2C \end{array}\!\!\!>\!\!C=N-B\!\!<\!\!\begin{array}{c} R \\ C(C_6H_5)_2 \end{array}$$
$$\qquad\qquad\qquad\qquad\qquad\qquad\qquad\qquad\qquad\qquad\qquad Br \quad\; Br$$

(32)

R = CH_3 or C_6H_5

Halogen bonded to boron or carbon of the CNB grouping of the molecule seems to be of similar activity. Reactions of monomeric iminoboranes with organothiols lead to C-S substituted iminoboranes which will be discussed later (c.f. Sect. VI).

Replacement of halogen in dimeric iminoboranes containing tetracoordin-ated boron is not so readily accomplished. For example, bis[(dichloromethyl-eneamino)dichloroborane] reacts with C_4H_9MgCl in nonetheric solvents to yield dimeric (dichloromethyleneamino)butylchloroborane, whereas C_4H_9Li replaces both boron-bonded chlorine atoms by butyl groups [25].

$$[Cl_2C=NBCl_2]_2 \quad \begin{array}{c} \xrightarrow[\text{i-octan}]{+4n\text{-}C_4H_9MgBr}} [Cl_2C=N-B(C_4H_9)Cl]_2 \quad (70\%) \\ \xrightarrow[\text{pentane}]{+4n\text{-}C_4H_9Li} [Cl_2C=N-B(C_4H_9)_2]_2 \quad (90\%) \end{array} \qquad (33)$$

Independent preparation of the same materials (by 1,2-addition of butyldi-chloroborane and dibutylchloroborane respectively [25]) to chlorocyanide con-firms the suggested structures.

No pure product has been isolated from the reaction of four molar equivalents of C_6H_5Li with $[(C_6H_5)_2C=NBBr_2]_2$ [47]. It is possible that this event is due to the use of an ethereal solution of phenyllithium, the ether being cleaved by the $>BBr_2$ groups.

Various attempts to replace boron-bonded chlorine by fluorine have been unsuccess-ful (see Sect. IV/2.). Substitution of halogen in dimeric iminoboranes by pseudohalogen groups has not been achieved with the notable exception of the azido group. Several di- and tetrazido bis(iminoboranes), (IX), have been synthesized by reacting dimeric chloro- or bromoiminoboranes with sodium azide in acetonitrile [25]. Even carbon-bonded halogen can be replaced by the N_3 group if a large excess of NaN_3 is used.

$R = N_3$ or CH_3
$X = N_3$ or CH_3
$Y = Cl$, Br or haloalkyl
$Z = Cl$, Br or N_3

(IX)

Dimeric (azido)iminoboranes are rather explosive materials.

Monomeric (trichloromethylbromomethyleneimino)methylbromoborane reacts exo-thermically with methylthiocyanate as shown by Eq. (34).

$$\begin{array}{c} \text{Cl}_3\text{C} \\ \text{Br} \end{array}\!\!C=N-B\!\!\begin{array}{c} \text{CH}_3 \\ \text{Br} \end{array} + 2\,CH_3SCN \longrightarrow \begin{array}{c} \text{Cl}_3\text{C} \\ \text{SCN} \end{array}\!\!C=N-B\!\!\begin{array}{c} \text{CH}_3 \\ \text{NCS} \end{array} + 2\,CH_3Br \qquad (34)$$

It is assumed that the NCS groups are bonded to the molecule via the N-atoms [27].

For a compilation of iminoboranes derived from haloboranes and organohaloboranes see Table 4. However, aldiminoboranes are listed in Table 3 and sulfur-containing compounds in Table 7. Bis- and tris(imino)boranes are listed in Table 6.

V. Iminoboranes Derived from Triorganoboranes

Only few examples are known for an addition reaction of trialkylboranes across a nitrile group according to Eq. (35) [2] (but c.f. Sect. III, Eq. (11)). The low reaction temperature required for this process is quite surprising.

$$(CH_2{=}CH{-}CH_2)_3B + RCN \xrightarrow[\text{i-pentane}]{20\,°C}$$

$$\xrightarrow{100\,°C}$$

$$(35)$$

$$R = CH_3, C_6H_5, CH_2{-}CH \qquad (X)$$

The structure of compound (X) was confirmed by its i.r. and n.m.r. spectrum. However, such a transformation as examplified by Eq. (35) is not consistent with the tendency to retain the iminoborane structure which is predominant even in compounds containing more mobile BH groups (c.f. Eq. (8) also c.f. Sect. II).

A number of fully organosubstituted iminoboranes has been obtained originating from diphenylketimine or its derivatives and triorganoboranes or diorganohaloboranes. In particular, monomeric (diarylmethyleneamino)diorganoboranes have been prepared in this manner. Four reaction routes have been used which correspond, in part, to procedures which have also been useful for the synthesis of compounds cited in Sects. III and IV [47].

$$
\left.
\begin{aligned}
&\text{a) } R_2^1C{=}N{-}Si(CH_3)_3 + R_2^2BX \\
&\text{b) } R_2^1C{=}NLi + R_2^2BX \\
&\text{c) } R_2^1C{=}NH + R_2^2BX \\
&\text{d) } R_2^1C{=}NH_2\,Cl + NaB(C_6H_5)_4
\end{aligned}
\right\} \rightarrow R_2^1C{=}N{-}BR_2^2 \quad (XI)
$$

R^1 and R^2 being aryl groups

The reaction product of (diphenylmethyleneimino)trimethylsilane with 2,2'-biphenylene-fluoroborane is an associated species as illustrated by (XII).

$$
\left[(C_6H_5)_2 C=N\dot{B} \underset{}{\bigcirc\bigcirc} \right]_n
$$

(XII)

(Diphenylmethyleneamino)dimenthylborane has been obtained from thermal dimethylation of diphenylketimine-trimethylborane (c.f. Eq. (8)). This compound was originally thought to be monomeric on the basis of mass spectroscopic evidence [40]. However, due to the infrared frequency of $\nu(C=N)$ observed at 1662 cm^{-1} and a relatively high m.p. of 173 °C it is now considered to be dimeric in the solid state [47].

For a listening of compounds mentioned in Sect. V see Table 5.

VI. Iminoboranes Derived from Organothioboranes

1. Addition of B—S, B—C, or B—Cl Bonds Across the C≡N Groups of Substitudes Nitriles

The first examples of sulfur-containing iminoboranes have been obtained by the 1,2-addition of the B—S bond of (alkylthio)diorganoboranes to acetonitrile [35].

$$
2\ R_2BSR' + 2\ CH_3CN \longrightarrow
\begin{array}{c}
CH_3 \underset{C}{\diagdown} SR' \\
\| \\
R \diagdown \underset{B}{} N \diagup R \\
R' \diagup B \underset{N}{} \diagdown R \\
\| \\
R'S \diagup C \diagdown CH_3
\end{array}
\tag{36}
$$

R = R = C$_3$H$_7$ or C$_4$H$_9$ or C$_6$H$_5$; R' = C$_2$H$_5$ or C$_4$H$_9$

Dimeric (methylthioalkylmethyleneamino)diorganoboranes produced in this reaction (mixed cis-trans forms) are in equilibrium with their monomers in hot chloroform solution.

Similar compounds can be prepared from organic thiocyanates and trialkylboranes [33] as illustrated in Eq. (37).

$$2 \text{ R}'-\text{S}-\text{C}\equiv\text{N} + 2 \text{ R}_3\text{B} \longrightarrow \quad \underset{\text{(XIII)}}{\begin{array}{c} \text{R}\diagdown_{\text{C}}\diagup^{\text{SR}'} \\ \| \\ \text{R}\diagdown_{\text{B}}\diagup^{\text{N}}\diagdown_{\text{B}}\diagup^{\text{R}} \\ \text{R} \quad \text{N} \quad \text{R} \\ \| \\ \text{R}'\text{S}\diagup^{\text{C}}\diagdown\text{R} \end{array}} \qquad (37)$$

$$R = i\text{--}C_3H_7; \quad R' = C_6H_5 \qquad\qquad\qquad (XIII)$$

Compound (XIII) exists as a mixture of monomeric and dimeric species at room temperature. The same situation is true for the reaction products from methyl- and isopropyl-thiocyanate, respectively, with triisopropylborane, whereas the iminoborane obtained from reacting methyl-thiocyanate with tri-n-butylborane is dimeric [33]. With the exception of (XIII) these compounds have not been characterized by analysis.

The reaction of tris(organothio)boranes with nitriles leads to B—S substituted iminoboranes. 1,2-addition of tris(methylthio)borane or tris(phenylthio) borane to trichloroacetonitrile yields the monomeric products (XIV) or (XV)[32].

$$\underset{(XIV)}{\begin{array}{c} Cl_3C\diagdown \qquad \diagup SCH_3 \\ C=N-B \\ CH_3S\diagup \qquad \diagdown SCH_3 \end{array}} \qquad\qquad \underset{(XV)}{\begin{array}{c} Cl_3C\diagdown \qquad \diagup SC_6H_5 \\ C=N-B \\ C_6H_5S\diagup \qquad \diagdown SC_6H_5 \end{array}}$$

Dimeric iminoboranes, on the other hand, are formed from fluoroacetonitrile and tris(organothio)boranes:

$$2 \text{ B(SR)}_3 + 2 \text{ FCH}_2\text{CN} \longrightarrow [\text{FCH}_2(\text{RS})\text{C}=\text{N}-\text{B(SR)}_2]_2 \qquad (38)$$

$$R = CH_3 \text{ or } C_6H_5$$

Dimeric(fluoromethylmethylthiomethyleneamino)bis(thiomethyl)borane monomerizes in CCl$_4$ solution [32].

Reactions of tris(organothio)boranes with α-halogenated nitriles are very slow (c.f. (II)). Analogous reactions of tris(organothio)boranes with acetonitrile could not be accomplished.

Monomeric iminoboranes containing only sulfur substituents have resulted from the thioboration of organic thiocyanates [32].

$$(CH_3S)_3B + R-S-C\equiv N \longrightarrow \underset{}{\begin{array}{c} RS\diagdown \qquad \diagup SCH_3 \\ C=N-B \\ CH_3S\diagup \qquad \diagdown SCH_3 \end{array}} \qquad (39)$$

$$R = CH_3 \text{ or } i\text{--}C_3H_7$$

Monomeric and dimeric species obtained from the reaction of alkylthio-cyanates with tris(phenylthio)borane have not been isolated in a pure state [32]. The reaction of (organothio)dichloroboranes with nitriles as illustrated by

Eq. (40) proceeds at extremely slow rate leading to the assumption that the reaction involves the B-S bond rather than the B-Cl bond. This event has been substantiated by ^{11}B n.m.r. spectroscopy of a dimeric monothioalkylborane (c.f. Sect. VIII.2) [33].

$$2\ CH_3SBCl_2 + 2\ RCN \longrightarrow \qquad\qquad\qquad\qquad (40)$$

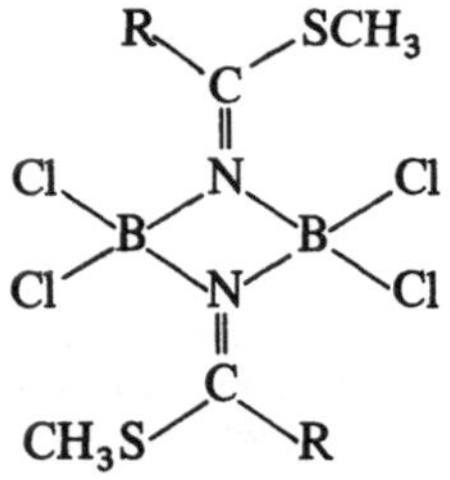

R = FCH_2, CH_3, C_2H_5, n-C_3H_7, i-C_3H_7, C_6H_5, or C_6F_5

Some of these compounds show unusual hydrolytic stability and $[C_4H_9S(C_6H_5)C=N-BCl_2]_2$ obtained in an analogous reaction from (butylthio)dichloroborane and benzonitrile has been recrystallized from aqueous acetone. The compound $[CH_3S(C_6F_5)C=N-BCl_2]_2$ can be sublimed in high vacuum without decomposition or any rearrangement in contrast to $[Br(C_6F_5)C=N-BBr_2]_2$ (c.f. Eq. (22)) [33].

Additional C–S substituted derivatives are formed by thermal rearrangements of phenylthiocyanate-trichloroborane and chloromethylthiocyanate-trichloroborane adducts respectively:

$$2\ R-S-C\equiv N + 2\ BCl_3 \longrightarrow 2\ RSCN:BCl_3 \xrightarrow[76\ °C]{CCl_4} \qquad\qquad (41)$$

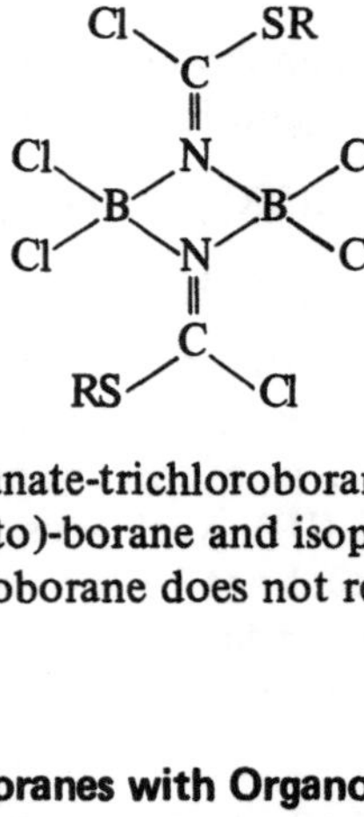

R = C_6H_5 or $ClCH_2$

However, thermal decomposition of isopropylthiocyanate-trichloroborane does not yield an iminoborane, rather, tris(thiocyanato)-borane and isopropylchloride are produced. Methylisothiocyanate-trichloroborane does not rearrange under comparable conditions [33].

2. Substitution Reactions on Halosubstituted Iminoboranes with Organothiols

When monomeric bromosubstituted iminoboranes are reacted with alkylthiols, adducts of thioester imides with haloboranes are formed [27].

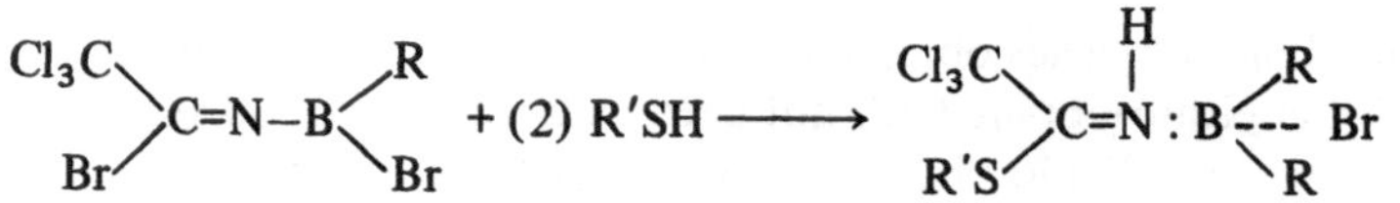

$$\underset{Br}{\overset{Cl_3C}{>}}C=N-B\underset{Br}{\overset{R}{<}} + (2)\ R'SH \longrightarrow \underset{R'S}{\overset{Cl_3C}{>}}C=N:B\overset{H}{---}Br \qquad (42)$$

54

The thioesterimide-haloboranes, upon thermal dehydrobromation, yield C–S substituted iminoboranes, whereas treatment with triethylamine leads to the rather unstable thioesterimides and triethylamine-borane adducts.

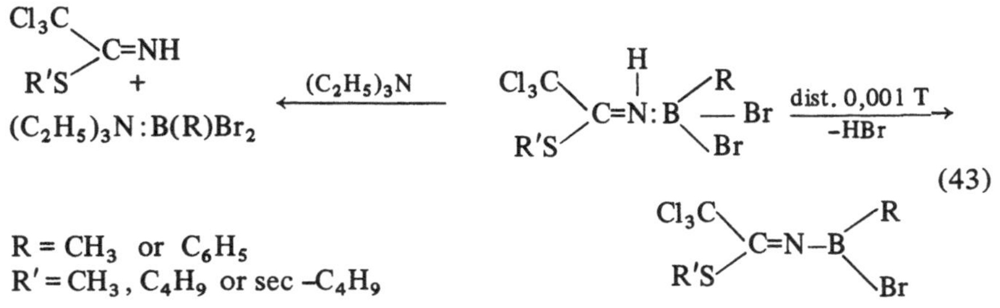

$$\underset{\substack{\text{Cl}_3\text{C} \\ \text{R}'\text{S}}}{\diagdown}\text{C=NH} + (\text{C}_2\text{H}_5)_3\text{N} : \text{B(R)Br}_2 \xleftarrow{(\text{C}_2\text{H}_5)_3\text{N}} \underset{\substack{\text{Cl}_3\text{C} \\ \text{R}'\text{S}}}{\diagdown}\text{C=N:B}\overset{\overset{\text{H}}{|}}{\underset{\text{Br}}{\diagup\text{R}}}\!\!-\!\text{Br} \xrightarrow[\text{–HBr}]{\text{dist. 0,001 T}}$$

$$\underset{\substack{\text{Cl}_3\text{C} \\ \text{R}'\text{S}}}{\diagdown}\text{C=N–B}\underset{\text{Br}}{\overset{\text{R}}{\diagup}}$$

(43)

R = CH$_3$ or C$_6$H$_5$

R' = CH$_3$, C$_4$H$_9$ or sec –C$_4$H$_9$

Analogous monomeric iminoboranes which are fully alkylated at the B-atoms have also been prepared [27]. Treatment of bromosubstituted iminoboranes with phenylthiol, however, also causes the B-bonded halogen to be replaced by thiophenyl groups [27] (Eq. 44).

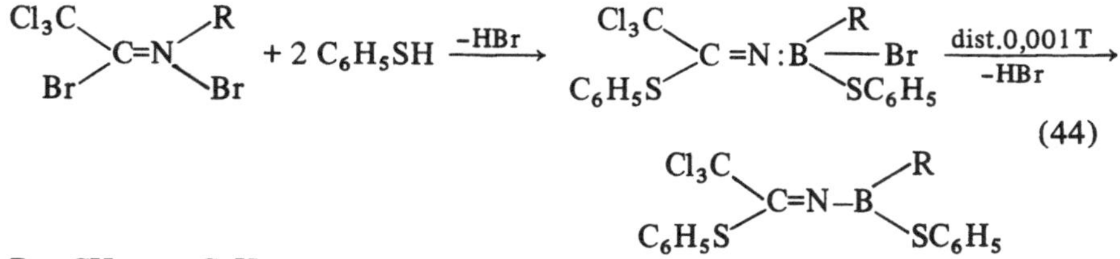

$$\underset{\text{Br}}{\overset{\text{Cl}_3\text{C}}{\diagdown}}\text{C=N}\underset{\text{Br}}{\overset{\text{R}}{\diagup}} + 2\,\text{C}_6\text{H}_5\text{SH} \xrightarrow{\text{–HBr}} \underset{\text{C}_6\text{H}_5\text{S}}{\overset{\text{Cl}_3\text{C}}{\diagdown}}\text{C=N:B}\underset{\text{SC}_6\text{H}_5}{\overset{\text{R}}{\diagup}}\!\!-\!\text{Br} \xrightarrow[\text{–HBr}]{\text{dist.0,001 T}}$$

$$\underset{\text{C}_6\text{H}_5\text{S}}{\overset{\text{Cl}_3\text{C}}{\diagdown}}\text{C=N–B}\underset{\text{SC}_6\text{H}_5}{\overset{\text{R}}{\diagup}}$$

(44)

R = CH$_3$ or C$_6$H$_5$

The reaction of alkylthiols with halosubstituted dimeric iminoboranes is another route for the preparation of some derivatives discribed in Sect. V. 1., but also leads to dimeric iminoboranes with only one organothio substituent [33].

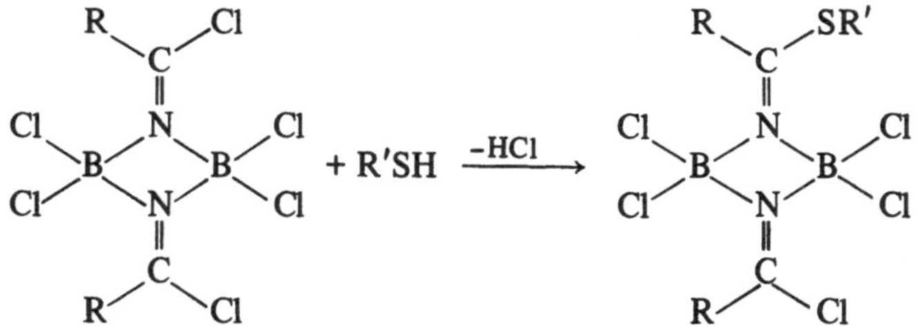

(45)

R = Cl or CH$_2$Cl; R' = CH$_3$ or n–C$_4$H$_9$

Dimeric iminoboranes with two alkylthio groups, (XVI), result when an excess of alkylthiol is used.

$$Y = CH_2F, Cl, CH_2Cl$$

(XVI)

Compounds of type (XVI) are practically insoluble in non-polar solvents or even in CH_3CN. They are only slowly hydrolyzed in boiling ethanolic NaOH solutions [33].

All sulfur-containing iminoboranes are compiled in Table 7.

VII. Imine-Borane Adducts

Like iminoboranes, imine-adducts of boron compounds have only recently been prepared and investigated. The adduct obtained from the reaction of diphenylketimine and diborane readily loses hydrogen, even at 20°, to form 1,3,5-diphenylmethylborazine [41]:

$$3 (C_6H_5)_2C=NH + 1,5 B_2H_6 \longrightarrow$$

$$3 (C_6H_5)_2C=N \overset{\overset{\displaystyle H}{|}}{:} BH_3 \xrightarrow{-3 H_2} [(C_6H_5)_2CHNBH]_3 \qquad (46)$$

Diphenylketimine forms an unstable solid adduct with trimethylborane, which, above 160 °C, slowly eliminates methane under formation of (diphenylmethyl-eneamino)methylborane [40].

$$2 (C_6H_5)_2C=NH + 2 B(CH_3)_3 \longrightarrow \qquad (47)$$

$$2 (C_6H_5)_2C=NH : B(CH_3)_3 \xrightarrow[-2 CH_4]{160-200\ °C} [(C_6H_5)_2 C=N-B(CH_3)_2]_2$$

The dissociation pressure of trimethylborane over the adduct is 23 Torr at room temperature and trimethylborane is readily pumped off in vacuo. Tri-ethylborane and triphenylborane did not afford adducts with diphenylketimine; also, the B-phenyl derivatives could not be prepared. However, the reaction between di(p-tolyl)methyleneimine hydrochloride and sodium tetraphenylborate produces an adduct which formed (di(p-tolyl)methyleneamino)diphenylborane at 230 °C [47].

$$(p\text{-}CH_3-C_6H_4)_2C=NH_2Cl + NaB(C_6H_5)_4 \xrightarrow[-C_6H_6]{-NaCl}$$

(48)

$$(p\text{-}CH_3-C_6H_4)_2C=NH:B(C_6H_5)_3 \xrightarrow[-C_6H_6]{230\,°C} (p\text{-}CH_3-C_6H_4)_2C=N-B(C_6H_5)_2$$

Solid adducts between $(C_6H_5)_2C=NH$ or $C_6H_5-CH=N-C_6H_5$ and trifluoroborane are known [41,43] and $(C_6H_5)_2C=NH:BF_3$ is sufficiently stable to withstand vacuum sublimation at 120 °C. Attempted preparation of corresponding BCl_3 derivatives led only to (diphenyl-methyleneimino)boranes [41]. Recently, additional imine-trifluoroborane adducts have been described [44]; they are listed in Table 8.

Dichloromethylenephenylamine and dichloromethyleneethylamine yield solid adducts [23] upon treatment with trichloro- or tribromoborane.

$$R-N=CCl_2 + BX_3 \longrightarrow Cl_2C=\overset{\displaystyle R}{\underset{\displaystyle |}{N}}:BX_3$$

(49)

$R = C_6H_5$ or C_2H_5; $X = Cl$ or Br

The adducts with BCl_3 can be sublimed without decomposition, whereas those with BBr_3 decompose in vacuo reverting to the starting materials. The adducts are extremely sensitive towards hydrolysis, the latter event leading to the free dichloromethyleneorganoamines and boric acid.

It is apparent that the B–N bond in imine-borane adducts is not very strong and is influenced by the donor and acceptor properties of the components as well as by steric factors. In the infrared spectra of the adducts the C=N stretching bands show only slight shifts as compared with those of the free imino derivatives [23,44].

Thioesterimide-haloborane adducts (c.f. Sect. VI.2, Eqs. (42–44)) can also be classified as imine-boranes. They are intermediated of low stability in the synthesis of monomeric sulfur substituted iminoboranes [27].

Imine-borane bonds which are incorporated into annular systems appear to be considerably more stable. A number of such compounds has been prepared by the reaction of nitriles with (cyclohexenylamino)boranes [7].

(50)

$R'' = CH_3$, C_6H_5 or C_3H_7; $R' = C_6H_5$, $i\text{-}C_4H_9$ or $i\text{-}C_5H_{11}$
$R = i\text{-}C_3H_7$ or $i\text{-}C_4H_9$

These coordinated heterocycles are not very sensitive towards atmospheric moisture. Upon hydrolysis in acidic medium diketonates, (e.g., acetylcyclohexanone) and diorganohydroxoboranes are obtained, which again form complexes of type (XVII).

(XVII)

For a listing of imine-borane adducts see Table 8.

VIII. Spectroscopic Studies on Iminoboranes

1. Infrared Spectra

Infrared spectroscopy is an excellent tool in iminoborane chemistry, which readily permits, to distinguish between iminoboranes and nitrile-borane adducts and to identify monomeric and dimeric forms of iminoboranes. This event is due to the fact that the νCN of CN multiple bonds absorbs outside the fingerprint region and can be considered to be a valuable group frequency even when mixed with other vibrational modes. In some cases other vibrations like NH, BH, B-halogen or B–S stretching modes are helpful for determining the structure of iminoboranes.

In the spectra of those nitrile-boranes having a CN bond order of about 3 the absorption due to the CN triple bond is observed near 2300 cm^{-1}, which is higher than the CN frequency of the free nitriles [19]. This event has been attributed to strict sp-hybridisation and to the nonexistence of canonical forms involving C=N double bonds in these adducts [9].

In iminoboranes the CN bond order is about two. The CN stretching mode in dimeric species is evidenced by an absorption between 1520 and 1700 cm^{-1}. Sulfursubstituted derivatives exhibit the CN band near the low frequency side of this region and CH$_3$ and H substituted species absorb near the high frequency side. Most dimeric iminoboranes show the C=N band around 1600 cm^{-1}. Some derivatives show a doublet which is also observed in derivatives containing the imine group (e.g. c.f. [15]). In most cases this event migth be caused by Fermi resonance with overtone or combination bands belonging to the same symmetry of vibrational type. Sometimes splitting is caused by unsymmetrical substitution, as in dimeric monothioalkylideneaminoboranes (c.f. Sect. VII.2.).

Monomeric iminoboranes exhibit a B–N bond order higher than unity due to $p_\pi - p_\pi$ bonding between nitrogen and three-coordinate boron. This event results in an allene-type structure as shown in (I) exhibiting its antisymmetric stretching vibration around 1800 cm^{-1}. This should have a predominant ν(CN) character, whereas in the symmetric mode of lower wavenumber the B-N character prevails. Nevertheless, referring to the 1800 cm^{-1} band as ν_{as}(CNB) appears to be reasonable from a spectroscopic point of view and this band (in the monomeric compounds) will be referred to as ν(C=N) through the present work in order to permit comparision with the (C=N) stretching mode in the dimeric species.

Due to the significance of the (C=N) stretching mode in iminoboranes, its frequency is included in the tabulations. Generally, the ν(C=N) frequency is affected by substituents on the $>$ C= N−B $<$ skeleton in the order given: H $>$ alk. $>$ Cl $>$ Br $>$ ar. $>$ S.

Typical spectra (4000 – 1250 cm^{-1}) of:
a nitrile-borane (CH$_3$C≡N : BCl$_3$, Fig. 1, p. 74)
a monomeric iminoborane (Br(CCl$_3$)C=N−B (CH$_3$)Br, Fig. 2, p. 74)
and a dimeric iminoborane [(Br(C$_6$F$_5$)C=N−BBr$_2$]$_2$, Fig. 3, p. 75)
are depicted.

In dimeric alkylidenamino-t-butylboranes (c.f. Eq. (9)), ν(BH) is found at 2390 cm^{-1} (derivatives of alkylnitriles) or at 2350 cm^{-1} derivatives of arylnitriles) [10]. Associated [Cl$_3$C−CH=N−BH$_2$]$_n$ has a BH stretching band near 2400 cm^{-1} [22], the CH stretch of dimeric $\left[\begin{smallmatrix} H \\ R \end{smallmatrix} \!\! >\!\! C=N-BR_2 \right]_2$ is observed between 3005 and 3010 cm^{-1}.

In some cases the antisymmetric stretch of the BX$_2$ group results in clear group frequencies. For example, in highly halogenated dimeric iminoboranes, ν_{as}BCl$_2$ absorbs between 850–900 cm^{-1} (doublet with 20 cm^{-1} boron isotope splitting) and ν_{as}BBr$_2$ absorbs between 787–807 cm^{-1} (doublet ~ 15 cm^{-1} boron isotop splitting) [23,31]. In sulfursubstituted dimeric iminoboranes, ν_{as}BCl$_2$ absorbs at slightly higher frequencies with the ^{11}BCl$_2$ branch around 920 cm^{-1} [33]. These assignments have been sustantiated by studies with ^{10}B-enriched compounds [29]. A D$_{2h}$ symmetry has been postulated for [Cl$_2$C=N=BCl$_2$]$_2$ on the basis of its infrared spectrum [21]. Azido groups bonded to dimeric iminoboranes have ν_{as}N$_3$ around 2140 cm^{-1} and ν_sN$_3$ around 1345 cm^{-1} [25]. Another vibration mode which results in group frequency absorptions is the ν_{as}(BX)$_3$ mode to be found in imine-borane adducts [27,40,44].

2. Nuclear Magnetic Resonance and Nuclear Quadrupole Resonance Spectroscopy

Boron-11 n.m.r. spectroscopy is a very valuable means for the study of iminoboranes since it permits to distinguish between monomers and dimers. The former species, which contain three-coordinated boron, absorb at lower filed [chemical shift δ about −40 ppm relative to external F$_3$BO(C$_2$H$_5$)$_2$] whereas the latter contain four-coordinated boron and absorb at higher fields (δ about −4 ppm). Monomer-dimer equlibria can be measured quantitatively by this method [39].

The boron-11 chemical shifts of some dimeric iminoboranes relative to $F_3B:O(C_2H_5)_2$ are listed in Table 1.

Table 1. *Boron-11 chemical shift data of dimeric iminoboranes relative to external tri-fluoroborane-diethyletherate*

Compounds		^{11}B ppm	Ref.
[RCH=N–B(t.but.)H]₂	doublet	+1,1 and –8,2	10)
[CF₃(Cl)C=N–BCl₂]₂		– 5,6	4)
[CF₃(Br)C=N–BBr₂]₂		+2,3	4)
[(CH₃)₃CH=N–BCl₂]₂		–4,5	6)
[(CH₃)₃CH=N–BBr₂]₂		–4,3	6)
[Cl₂C=N–BCl₂]₂		– 5,0	21)
[CH₃S(C₆H₅)C=N–BCl₂]₂		– 3,2	33)
[C₄H₉S(C₆H₅)C=N–BCl₂]₂		– 4,3	33)

$$\begin{array}{c} Cl_2 \\ C_4H_9S \\ \diagdown C=N \diagup B \diagdown N=C \diagup Cl \\ Cl \diagup \diagdown B \diagdown \diagup Cl \\ Cl_2 \end{array}$$ – 4,1 (sharp) 33)

The boron-11 n.m.r. spectrum of the compound
$[(CH_3)_3CCH=N–B(C_4H_9)_2]_n$ exhibits a strong signal at –7,4 ppm and a weak one at –38,8 ppm. At 120 °C the signal intensities reverse thus indicating that the dimer reverts to the monomer with increasing temperature [6]. The chemical shifts of imine-borane derivatives (c.f. Eq. (5)) are observed near +1,0 ppm [7].

Proton magnetic resonance spectroscopy has been used in many cases [6,7, 17,33,40,47] in order to characterize iminoboranes. These spectra are of particular value for distinguishing between possible stereoisomers [6,17,47]. On the basis of ^{19}F n.m.r. it has been established that the compound $[CF_3CX=N–BX_2]_2$ (X = Cl, Br) exists as mixtures of isomers, since the resonance signal appeared as a doublet [4].

Clorine-35 nuclear quadrupole resonance studies of complexes of trichloroborane [48] show that the resonance frequencies of the complexed BCl_3 group are constant within about ± 1 Mcs centered around 21 Mcs and thus are not sensitive to the nature of the fourth ligand. Regarding to the symmetry of the complex one, two, or three distinct ^{35}Cl n.q.r. frequencies will appear in the 21 Mcs region. The reaction product of BCl_3 with trichloroacetonitrile, however, has six separate ^{35}Cl n.q.r. frequencies — two a closely spaced doublet in the 23 Mcs region, three in the neighborhood of 40 Mcs (belonging to the CCl_3 group) and one at 37,8 Mcs. Hence, one of the chlorine atoms originally attached to boron must have migrated to a more electronegative atom, which confirms an iminoborane structure.

Simlilar conclusions are indicated for the reaction product obtained from dichloro-acetonitrile and trichloroborane. For the product obtained from chloroacetonitrile and trichloroborane, however, a nitrile-borane structure is found which, according to the authors [48], might depend critically on the reaction conditions.

3. Mass Spectrocopic Data

Mass spectra have been used by several authors to elucidate iminoborane structures. The spectra give valuable information through fragmentation of the molecules, but should be used with caution in order to avoid erroneous conclusions. In general, the boron and halogen containing fragments are easily recognized by their isotope patterns [1,3,11].

In the mass spectra of iminoboranes the parent molecular peak is often of rather low intensity, or cannot be detected at all, particularly in the mass spectra of halogenated iminoboranes. In the spectra of the latter, fragments formed by loss of halogen atoms give very strong peaks. Dimeric iminoboranes, which tend to give monomers in the vapor phase might exhibit spectra without peaks at m/q values higher than the molecular weight of the monomer [40,47]. Group transfer from carbon to boron has been observed [47]. Some compounds for which detailed information about their mass spectrometric fragmentation is available are listed in Table 2.

Table 2. *Iminoboranes for which details of m.s. fragmentation have been published*

Compound	Ref.	Appearance of parent ion
$[Cl_2C=N-BCl_2]_2$	21)	−
$[(CH_3)_3CCH=N-BCl_2]_2$	6)	weak
$[(C_6H_5)_2C=N-BCl_2]_2$	13)	−
$[(C_6H_5)_2C=N-BBr_2]_2$	13)	−
$[(C_6H_5)_2C=N-BF_2]_n$	13)	−
$[(C_6H_5)_2C=N-B(Cl)C_6H_5]_2$	47)	−
$[(C_6H_5)_2C=N-B(CH_3)_2]_2$	40)	−
$(CH_3)_2C=N-B(C_6H_5)_2$	37)	−
$(C_6H_5)_2C=N-B(mesit.)_2$	47)	+
$C_6H_5CH=N-B(mesit.)_2$	47)	+
$[C_6H_5(CH_3S)C=N-BCl_2]_2$	33)	−

IX. Listing of Compounds

A listening of iminoboranes and imine-boranes is complied in Tables 3 – 8; the particular arrangement is somewhat arbitrary. In Table 3 all aldiminoboranes are listed regardless of other substituents. Table 4 depicts those fully organic substituted iminoboranes which are obtained from halogenated iminoboranes by substitution reactions. In Table 7 all sulfur-containing iminoboranes are listed. The sequence of compounds within the tables is based on increasing formula weigth for the monomer (XVIII)

$$\left[\begin{matrix} Y \\ Z \end{matrix} \diagup C = N - B \diagup \begin{matrix} R \\ X \end{matrix} \right]_n \qquad (XVIII)$$

in case of the 1,2-addition Y always represents the nitrile substituent, which is therefore easily recognized. Z indicates the group migrating from boron.
The column "methods of preparation" listing numbered equations or formulae should assist to locate relevant sections of the text. Superscripts used in the tables:

a: (partial) transformation into the corresponding nitrile-borane upon sublimation

b: in equilibrium with corresponding nitrile-borane at room temp.

c: forms the monomeric form upon sublimation at 11 mm

d: monomerizes in solution

e: monomeric in gas-phase

f: slow decomposition at 20 °C

g: slow decomposition at 0 °C

h: melting point of corresponding nitrile-borane adduct

l: liquid (at room temperature)

o: not isolated

s: solid (at room temperature)

In cases where temperature ranges for melting (decomposition) and (or) sublimation or boiling points are listed in the original literature, only the lowest temperature is cited in the following tables.

Table 3. *Iminoboranes derived from monoborine and organoboranes (aldimino-boranes)*

Y	Z	R	X	n	Methods of preparation acc. to	m.p. (°C) [dec.]	subl. p. or b.p. (°C/mm)	i.r. ν(CN) cm^{-1}	Ref.	Remarks
CH_3	H	CH_3	CH_3	2	(10)	$-5/76^+$	$20/1,8\,(0,6)^+$	1698	17)	+cis-trans
CH_3	H	H	$(CH_3)_3C$	2	(9)	75	72/0,8	1660	10)	
CH_3	H	C_2H_5	C_2H_5	2	(10)	-18	vac.	1692	17)	
C_2H_5	H	H	$(CH_3)_3C$	2	(9)	79	84/0,8	1660	10)	
$n\text{-}C_3H_7$	H	H	$(CH_3)_3C$	2	(9)		86/0,1	1660	10)	
$i\text{-}C_3H_7$	H	H	$(CH_3)_3C$	2	(9)	57	84/0,4	1660	10)	
CCl_3	H	H	H	2	(13)	116	110/vac.	1700*	16)	*22)
$(CH_3)_3C$	H	Cl	Cl	2	(31)	176	125/0,01	1684	6)	
C_6H_5	H	H	$(CH_3)_3C$	2	(9)	141		1640	10)	191ˣ
$4\text{-}CH_3\text{-}C_6H_4$	H	H	$(CH_3)_3C$	2	(9)	146		1640	10)	187ˣ
$3\text{-}CH_3\text{-}C_6H_4$	H	H	$(CH_3)_3C$	2	(9)	124		1640	10)	
$4\text{-}F\text{-}C_6H_4$	H	H	$(CH_3)_3C$	2	(9)	137		1640	10)	180ˣ
$4\text{-}CH_3O\text{-}C_6H_5$	H	H	$(CH_3)_3C$	2	(9)	145		1640	10)	185ˣ
$4\text{-}Cl\text{-}C_6H_4$	H	H	$(CH_3)_3C$	2	(9)	138		1640	10)	190ˣ
$(CH_3)_3C$	H	C_4H_9	C_4H_9	2	(11)	74	70/0,3	1674	5,6)	x = m.p. of
C_6H_5	H	C_6H_5	Cl	2	(27)	222		1639	47)	isomer
$(CH_3)_3C$	H	Br	Br	2	(31)	207	135/0,01	1662	6)	(structure
C_6H_5	H	C_6H_5	C_6H_5	2	(27)	122		1643	47)	unknown)
H	Br	Br	Br	1	(18)		65/11	1820	24)	
C_6H_5	H	mes.	mes.	1	(27)	97		1813	47)	

Table 4. *Iminoboranes derived from halogeno- and organohalogeno-boranes*

Substituents (Formula (XVIII))				n	Method(s) of preparation acc. to	m.p. (°C) [dec.]	subl.p. or b. p. (°C/mm)	i.r. ν(CN) cm^{-1}	Ref.
Y	Z	R	X						
CH_2F	N_3	CH_3	CH_3	2	(IX)			1665+1646	25)
CH_2Cl	N_3	CH_3	CH_3	2	(IX)			1680	25)
CH_2F	Cl	Cl	Cl	2	(2) (4)			1650	31)
Cl	Cl	Cl	Cl	2	(2) (7) (15)	[150]	130/0,001	1648+1602	21,23,30)
CH_2F	Br	CH_3	CH_3	2	(2) (4)	[186]	80/0,001	1675	31)
Br	Cl	CH_3	CH_3	2^c	(2) (19)	[138]	130/0,001	1648	24)
CH_3OCH_2	Cl	Cl	Cl	2	(2) (24)	100	120/11^c	1657+1598	33)
Cl	Cl	N_3	N_3	2	(IX)	60^f		1588	25)
CH_2Cl	Cl	Cl	Cl	2	(2) (4)	[120]	60/0,001^a	1648+[2342]	31,48)
CH_2Cl	Br	CH_3	CH_3	2+1	(2) (4)	f		1668+(1820w)	31)
Cl	Cl	Cl	C_4H_9	2	(2) (33)	125	105/0,001	1640	25)
CH_2Cl	Cl	N_3	N_3	2	(IX)			1615	25)
CH_2F	Br	N_3	CH_3	2	(IX)			1672	25)
CF_3	Cl	Cl	Cl	2	(2) (14)	80	25/0,001	1663+[1625sh]	4)
Cl	Cl	Cl	C_6H_5	2	(2) (16)	170	135/0,001	1635	23)
Cl	Cl	C_4H_9	C_4H_9	2	(2) (33)	1		1645	25)
Br	Cl	Cl	Cl	2	(2) (16)	[140]	95/0,001	1615	23)
CH_2Cl	Br	CH_3	CH_3	2	(IX)			1645	25)
Br	Br	CH_3	CH_3	2+1	(2)	90	70/0,001	1635+[1840w]	24)
$CHCl_2$	Cl	Cl	CH_3	2	(2)	[120]	110/12	1650	31,48)
C_6H_5	C_6H_5 F	Cl	F	n	(3)(4)(26)(27) (VII)	286		1620	13,46)

Table 4 (continued)

Substituents (Formula (XVIII))				n	Method(s) of preparation acc. to	m.p. (°C) [dec.]	supl. p. or b. p. (°C/mm)	i. r. ν(CN) cm^{-1}	Ref.
Y	Z	R	X						
F	Br	Br	CH₃	1+2	(2) (18)		75/0,001	1815+1785+1630	26)
CHCl₂	Br	CH₃	CH₃	2	(2)	[120]	70/0,001ᵃ	1662	31)
CH₂F	Br	N₃	N₃	2	(IX)			1620	25)
CH₂Br	Cl	Cl	Cl	2	(2)	[110]	120/11ᵃ	1646	31)
CH₂Br	Br	CH₃	CH₃	2	(2)	145	95/0,001	1640	31)
C₆H₅	C₆H₅	C₂H₅	C₂H₅	(1)				1793	12)
CCl₃	Br	CH₃	CH₃	1	(2)	1	105/11	1840+[1597m]	24)
Br	Cl	Cl	C₄H₉	2ᵇ	(2)		100/0,001	2285+1620	26)
CH₂F	Br	Br	CH₃	2	(2)	132		1665	31)
CH₂Cl	Br	N₃	N₃	2	(IX)			1625	25)
CH₂Cl₂	Br	N₃	CH₃	2	(IX)			1635	25)
CCl₃	Cl	Cl	Cl	2ᵈ	(2)	79	70/0,001	1640	24,48)
C₂F₅	Cl	Cl	Cl	2	(2)	80	90/0,001	1660	26)
CH₂Cl	Br	Br	CH₃	2	(2)	[105]		1657	31)
C₆H₅	C₆H₅	Cl	C₆H₅	2	(3) (4) (27) (VII)	[305]		1590	12,13,45)
Br	Cl	Cl	Br	2	(2)	160	140/0,001	1630	23)
Cl	Cl	Br	Br	2	(7)	[160]	95/0,001	1655sh+1605	21)
CH₂Br	Br	CH₃	CH₃	2	(IX)			1645	25)
Br	Br	N₃	N₃	2	(IX)			1582	25)
CH₂I	Cl	Cl	Cl	2ᵇ	(2)			1580+2330ᵇ	31)
C₆H₅	C₆H₅	Cl	C₆H₅	n(2)	(4) (27)	237		1620+1587	12,16,47)

65

Table 4 (continued)

Substituents (Formula XVIII))				n	Method(s) of preparation acc. to	m.p. (°C) [dec.]	supl. p. or b.p. (°C/mm)	i.r. ν(CN) cm^{-1}	Ref.
Y	Z	R	X						
F	Br	Br	Br	1	(2) (18)	f	80/0,001	1815+1785	26)
C_6H_5	C_6H_5	Br	$C_6H_4O_2$	2	(27) (VIII)	210	110/0,001	1624	46,47)
CH_2Br	Br	Br	CH_3	2	(2)	[147]	80/0,001	1655	31)
Cl	Cl	Cl	C_6F_5	2	(2) (21)	140	170/0,001	1640+1600	26)
CH_2F	Br	Br	Br	2	(2)	[190]	100/0,001	1631	31)
Cl	Br	Br	Br	2	(2) (15)	195		1603	23)
$CHBr_2$	Cl	Cl	Cl	2	(2)	[140]		1643	31)
$CHCl_2$	Br	Br	CH_3	2	(2)	[160]		1650	31)
$CHBr_2$	Br	CH_3	CH_3	2	(2)	134	85/12	1644	31)
CH_2Cl	Cl	Cl	C_6F_5	2^b	(2) (22)	101	80/0,001	2260+1650	26)
CH_2Cl	Br	Br	Br	2	(2)	[145]		1632	31)
CF_3	CF_3	C_6H_5	C_6H_5	1	(6)	2	110/22	1839	37)
CCl_3	Br	Br	CH_3	1	(2)	2	70/12	1835	24)
C_2F_5	Br	Br	CH_3	2+1	(2)	75	80/0,001	1850+1645	26)
CF_3	Br	Br	Br	2	(2) (14)	162	100/0,001	1645[+1693sh]	4)
$CHBr_2$	Br	N_3	CH_3	2	(IX)			1645	25)
C_6H_5	C_6H_5	Br	Br	2	(3) (4) (25) (27)	189		1586+1560	12,13,46)
C_6F_5	Br	N_3	CH_3	2	(IX)			1665+1645	25)
Br	Cl	Cl	C_6F_5	2^b	(2) (22)	255	130/0,001	2300+1620	26)
Br	Br	Br	Br	2	(2) (15)	[185]	100/0,001	1600	23)
$CHCl_2$	Br	Br	Br	2	(2)	[150]	90/001^a	1610	31)
CH_2Br	Br	Br	Br	2	(2)	[145]		1630	31)
CCl_3	Br	C_6H_5	C_6H_5	1	(2)	1		1840	24)

Table 4 (continued)

Substituents (Formula XVIII))				n	Method(s) of preparation acc. to	m.p. (°C) [dec.]	supl.p. or b.p. (°C/mm)	i.r. ν(CN) cm^{-1}	Ref.
Y	Z	R	X						
C_6F_5	Br	Br	CH_3	2	(2)(20)		80/0,001[a]	1655	24)
$CHBr_2$	Br	Br	CH_3	2	(2)	[180]	110/0,001	1634	31)
$4-CN-C_6F_4$	Br	Br	CH_3	2	(2)	[180]		1665+1645	26)
C_5H_4NHCl	Br	Br	Br	2	(2)(17)	[130]		1650+1623	26)
CCl_3	Br	Br	C_6H_5	1	(2)	—	185/11	1842	24)
CCl_3	Cl	Cl	C_6F_5	1+2	(2)	—	80/10	1870+1650	26)
CBr_3	Cl	Cl	Cl	2	(2)	[140]	110/0,001	1623	31)
CCl_3	Br	Br	Br	2[d]	(2)	[120]	75/0,001[e]	1610	24)
C_2F_5	Br	Br	Br	2	(2)	114	110/0,001	1638	26)
CBr_3	Br	CH_3	CH_3	1+2	(2)	1+s		1820+(1650w)	31)
CH_2I	Br	Br	Br	2[b]	(2)	f		2305+1590	31)
CCl_3	$(C_6H_5)_2CBr$	CH_3	CH_3	1	(32)	—		1815	28)
C_6F_5	Br	Br	C_6H_5	2	(2)(20)	162[h]	100/0,005[a]	1650	24)
C_6F_5	Br	Br	Br	2	(2)(20)	[200]	100/0,001	1645	24)
C_6H_5	C_6H_5	I	I	2	(3)(25)	1+s		1602+1564	13)
$CHBr_2$	Br	Br	Br	2+1	(2)			1625+1815	31)
$4-CN-C_6F_4$	Br	Br	Br	2	(2)	200		1658+1650	26)
CBr_3	Cl	Cl	Cl	2	(2)	[140]	110/0,001	1623	31)
CH_2Cl	I	I	I	2[b]	(2)	f		1580[+2310w]	31)
$CHCl_2$	I	I	I	2	(2)	f		1576	31)
CBr_3	Br	Br	Br	2+1	(2)	[80]		1575+[1790w]	31)
CCl_3	I	I	I	1	(2)	g		1830	31)
CCl_3	$(C_6H_5)_2Br$	C_6H_5	C_6H_5	1	(32)	40		1820	28)

67

Table 4 (continued)

Substituents (Formula (XVIII))				n	Method(s) of preparation acc. to	m.p. (°C) [dec.]	supl. p. or b. p. (°C/mm)	i.r. ν(CN) cm⁻¹	Ref.
Y	Z	R	X						
Br₂(CH₃)B : 4-CN–C₆F₄									
Br	Br	CH₃	CH₃	2	(2)	[185]		1665+1645	26)
CCl₃	(C₆H₅)₂CBr	(C₆H₅)₂CBr	CH₃	1	(32)	[100]		1825	28)
Br₃B : 4-CN–C₆F₄									
Br	Br	Br	Br	2	(2)	[230]		1660+1630	26)
CCl₃	(C₆H₅)₂Br	(C₆H₅)₂CBr	C₆H₅	1	(32)	[130]		1815	28)

Table 5. *Iminoboranes derived from tris(organo)boranes*

Substituents (Formula (XVIII))				n	Method(s) of preparation acc. to	m.p. (°C) [dec.]	subl. p. or b.p. (°C/mm)	i. r. ν(CN) cm⁻¹	Ref.
Y	Z	R	X						
CH₃	C₃H₅	C₃H₅	C₃H₅	2	(35)				2)
C₃H₅	C₃H₅	C₃H₅	C₃H₅	2	(35)				2)
C₆H₅	C₆H₅	CH₃	CH₃	2	(8)			1662	12,40)
C₆H₅	C₃H₅	C₃H₅	C₃H₅	2	(35)				2)
2,2'(C₆H₄)₂		C₆H₅	C₆H₅	1	(XIa) (4)	105		1776	47)
C₆H₅	C₆H₅	2,2'(C₆H₄)₂		2	(XII)	128		1616	47)
C₆H₅	C₆H₅	C₆H₅	C₆H₅	1	(3) (4) (XII c,d)	295	140/0,001	1736	12,45–47)
4–CH₃–C₆H₅	4–CH₃–C₆H₅	C₆H₅	C₆H₅	1	(3) (XII c,d) (48)	143	140/0,1	1775	47)
4–Cl–C₆H₄	4–Cl–C₆H₄	C₆H₅	C₆H₅	1	(XII c)	119		1792	47)
4–Br–C₆H₄	C₆H₅	C₆H₅	C₆H₅	1	(XII c)	1	140/0,1	1792	47)
C₆H₅	C₆H₅	mes.	mes.	1	(3) (4) (XII a,b)	168		1792	12,46,47)

Table 6. *Bis- and tris(imino)boranes*

	Method(s) of preparation acc. to	m.p. (°C) [dec.]	subl. p. or b.p. (°C/mm)	i.r. ν(CN) cm^{-1}	Ref.
B[N=C(CF₃)₂]₃					37)
C₆H₅B[N=C(C₆H₅)₂]₂	(3) (4) (28)	127		1672	12,46)
B[N=C(C₆H₅)₂]₃	(3) (4) (29) (30)			1650	12,46,47)

Table 7. *Iminoboranes derived from organothioboranes*

| Substituents (Formula (XVIII)) | | | | n | Method(s) of preparation acc. to | m. p. (°C) [dec.] | subl. p. or b.p. (°C/mm) | i.r. ν(CN) cm^{-1} | Ref. |
Y	Z	X	R						
CH₃	CH₃S	Cl	Cl	2	(XVI)	[170]		1562	33)
C₂H₅	CH₃S	Cl	Cl	2	(40)	[220]		1562	33)
CH₂F	CH₃S	Cl	Cl	2	(40) (XVI)	[165]		1566	33)
Cl	CH₃S	Cl	Cl	2	(XVI)	[170]		1529	33)
n–C₃H₇	CH₃S	Cl	Cl	2	(40)	[170]		1540	33)
i–C₃H₇	CH₃S	Cl	Cl	2	(40)	[150]		1547+1520	33)
CH₃	C₂H₅S	C₃H₇	C₃H₇	2^d	(36)	96			35)
CH₂Cl	CH₃S	Cl	Cl	2	(XVI)	[140]		1562	33)
CH₂F	CH₃S	CH₃S	CH₃S	2^d	(38)	[84]		1581	32)
Cl	ClCH₂S	Cl	Cl	2	(41)	[162]		1551	33)
CH₃S	CH₃S	CH₃S	CH₃S	1	(39)	1		1752+1648	32)
CH₃	C₄H₉S	C₃H₇	C₃H₇	2^d	(36)	89		1629	35)

Table 7 (continued)

| Substituents (Formula (XVIII)) | | | | n | Method(s) of preparation acc. to | m. p. (°C) [dec.] | subl. p. or b. p. (°C/mm) | i. r. ν(CN) cm⁻¹ | Ref. |
Y	Z	R	X						
i-C₃H₇	CH₃S	Cl	Cl	2	(40)	[150]		1547+1520	33)
CH₃	C₄H₉S	C₄H₉	C₄H₉	2ᵈ	(36)	86			35)
Cl	C₆H₅S	Cl	Cl	2	(41)	[193]		1545	33)
i-C₃H₇S	CH₃S	CH₃S	CH₃S	1+(2)	(39)	—		1762+1657	32)
CCl₃	CH₃S	Cl	Cl	1	(38)	—		1822+1768	33)
CCl₃	n–C₄H₉S	CH₃	CH₃	1	(43)	—	70/0,001	1805	27)
CCl₃	sec.–C₄H₉S	CH₃	CH₃	1	(43)	—	60/0,001	1810	27)
C₆H₅	n–C₄H₉S	Cl	Cl	2	(40)	[230]		1558	33)
i-C₃H₇	C₆H₅S	i–C₃H₇	i–C₃H₇	1+2	(37)	—	75/0,001		33)
CCl₃	NCS	NCS	CH₃	(1)	(34)	—		1610+1540	27)
CH₃	C₄H₉S	C₆H₅	C₆H₅	2⁽ᶠ⁾	(36)	120			35)
CCl₃	CH₃S	CH₃S	CH₃s	1	(XIV)	—		1790+1755	32)
CCl₃	CH₃S	CH₃	CH₃	1	(43)	—	50/0,001	1810	27)
C₆F₅	CH₃S	Cl	Cl	2	(40)	[250]		1651+1569	33)
CCl₃	n–C₄H₉S	Br	CH₃	1	(43)	—	78/0,001	1800	27)
CCl₃	sec.–C₄H₉S	Br	CH₃	1	(43)	—	77/0,001	1800	27)
CCl₃	CH₃S	Br	C₆H₅	1	(43)	—	95/0,001	1810	27)
CCl₃	C₆H₅S	CH₃	CH₃	1	(44)	—	110/0,001	1805	27)
CH₂F	C₆H₅S	C₆H₅S	C₆H₅S	2	(39)	126		1578	32)
CCl₃	n–C₄H₉S	Br	C₆H₅	1	(43)	—	100/0,001	1803	27)

Table 7 (continued)

Substituents (Formula (XVIII))				n	Method(s) of preparation acc. to	m.p. (°C) [dec.]	subl. p. ν(CN) (°C/mm)	i.r. cm⁻¹	Ref.
Y	Z	R	X						
CCl₃	sec.-C₄H₉S	Br	C₆H₅	1	(43)	1	100/0,001	1805	27)
CH₃S	C₆H₅S	C₆H₅S	C₆H₅S	1+2	(39)	1°		1751+1640	32)
i-C₃H₇S	C₆H₅S	C₆H₅S	C₆H₅S	2	(39)	1°		1748+1634	32)
CCl₃	C₆H₅S	C₆H₅S	C₆H₅	1	(44)	1	155/0,001	1805	27)
CCl₃	C₆H₅S	C₆H₅S	C₆H₅S	1	(XV)	1		1801	32)

Dimeric mono(organothio)iminoboranes

	Method	m.p.	subl. p.	i.r.	Ref.
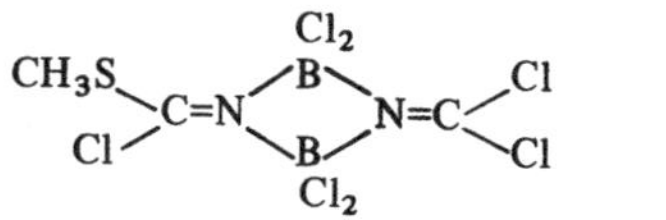	(45)	[150]	120/0,001	1678+1602+ 1558	33)
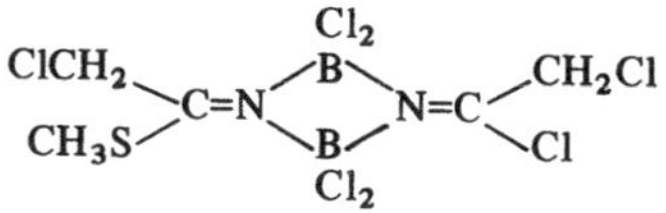	(45)	[129]		1675+1611+ 1559	33)
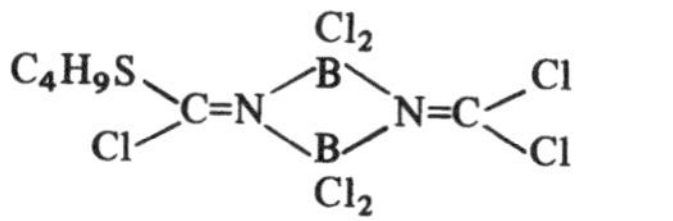	(45)	114		1671+1622+ 1552	33)

71

Table 8. *Imine-borane adducts*

Compound	Method(s) of preparation acc. to	m. p. (°C) [dec.]	subl. p. or b. p. (°C/mm)	i. r. ν(CN) cm^{-1}	Ref.
Aldimine-borane adducts					
$C_6H_5CH=NCH_3:BF_3$		126		1712	44)
$C_6H_5CH=NC_6H_5:BF_3$		154		1673	43,44)
Ketimine-borane adducts					
$(C_6H_5)_2C=NH:BH_3$	(46)	[90]		1620	41)
$(C_6H_5)_2C=NH:B(CH_3)_3$	(47)			1604	40)
$(4-C_6H_3-C_6H_4)_2C=NH:B(C_6H_5)_3$	(48)	171		1594	47)
$[(CH_3)_3C]_2C=NH:BF_3$		88		1672	44)
$4-CH_3-C_6H_4[(CH_3)_3C]C=NH:BF_3$		118		1666	44)
$(C_6H_5)_2C=NH:BF_3$		205		1628	41,44)
$(4-CH_3-C_6H_4)_2C=NH:BF_3$		[200]		1626	44)
$(4-Cl-C_6H_4)_2C=NH:BF_3$		76		1633	44)
$4-Br-C_6H_4(C_6H_5)C=NH:BF_3$		185		1629	44)
$(C_6H_5)_2C=NCH_3:BF_3$		111		1661	44)
$(C_6H_5)_2C=NC_6H_5:BF_3$		232		1621	44)
Dichloromethyleneamine-borane adducts					
$C_2H_5N=CCl_2:BCl_3$	(49)	[92]	subl/11		23)
$C_6H_5N=CCl_2:BCl_3$	(49)	132	subl/760		23)
$C_2H_5N=CCl_2:BBr_3$	(49)	[70]			23)
$C_6H_5N=CCl_2:BBr_3$	(49)	[88]			23)

Table 8 (continued)

Thioesterimide-Borane adducts

Compound		Method(s) of preparation acc. to	m.p. (°C) [dec.]	subl. p. or b.p. (°C/mm)	i.r. ν(CN) cm⁻¹	Ref.
Cl_3C–C=NH:B<$\genfrac{}{}{0pt}{}{CH_3}{Br}$ / RS–Br	R = CH₃	(42)	[110]		1598	27)
	R = nC₄H₉	(42)	[1]		1602	27)
	R = sec C₄H₉	(42)	[1]		1592	27)
Cl_3C–C=NH:B<$\genfrac{}{}{0pt}{}{C_6H_5}{Br}$ / RS–Br	R = CH₃	(42)	1+s		1640	27)
	R = nC₄H₉	(42)	ready	decomp.		27)
	R = sec C₄H₉	(42)	ready	decomp.		27)
Cl_3C–C=NH:B<$\genfrac{}{}{0pt}{}{CH_3}{CH_3}$ / RS–Br	R = nC₄H₉	(42)	[75]		1590	27)
	R = sec C₄H₉	(42)	[85]		1598	27)
Cl_3C–C=NH:B<$\genfrac{}{}{0pt}{}{R}{SC_6H_5}$ / C_6H_5S	R = CH₃	(44)	1		1605	27)
	R = C₆H₅	(44)	1	ready decomp.		27)
(bicyclic structure, X–C=NH, N–$B(nC_3H_7)_2$, N–R)	X = CH₃ R = C₆H₅	(51)	106	175/1,5	1520+1595	7)
	X = C₆H₅ R = i.C₄H₉	(51)		174/0,3	1520+1595	7)
	X = C₃H₇ R = C₆H₅	(51)		175/0,8	1520+1595	7)
(bicyclic structure, X–C=NH, N–$B(nC_4H_9)_2$, N–R)	X = CH₃ R = C₆H₅	(51)	63	185/1	1520+1595	7)
	X = CH₃ R = n-C₅H₁₁	(51)		174/1,5	1520+1595	7)
	X = C₆H₅ R = C₆H₅	(51)	51	210/0,5	1520+1595	7)

A. Meller

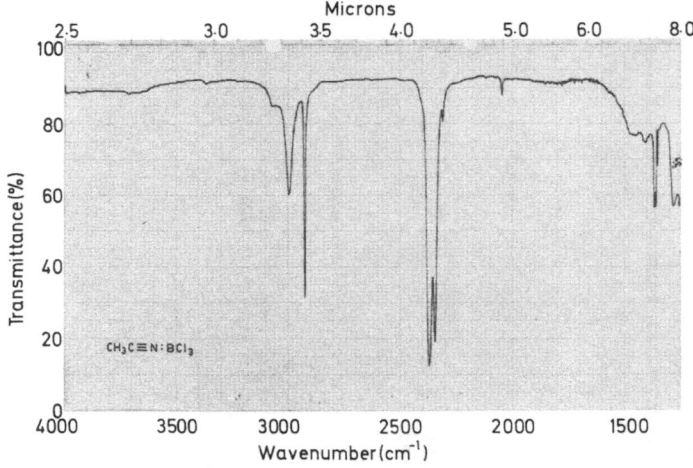

Fig. 1. I.R.-Spectrum of Acetonitrile-trichloroborane [mull in $(ClCF=CH_2)_n$]

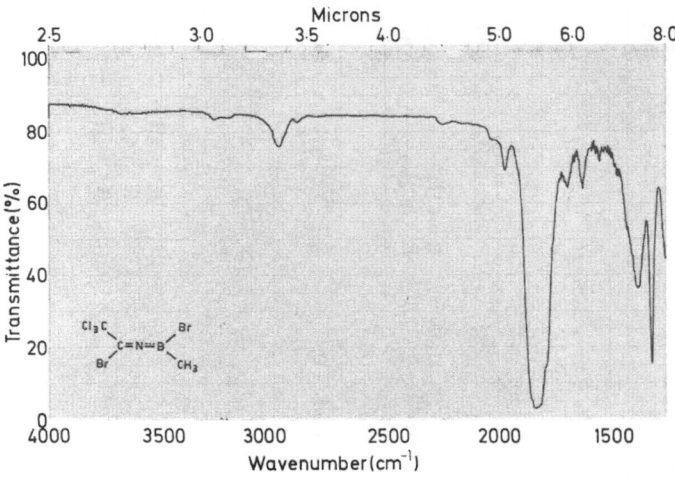

Fig. 2. I.R.-Spektrum of Trichloromethylbromomethyleneaminodibromoborane [solution in CCl_4]

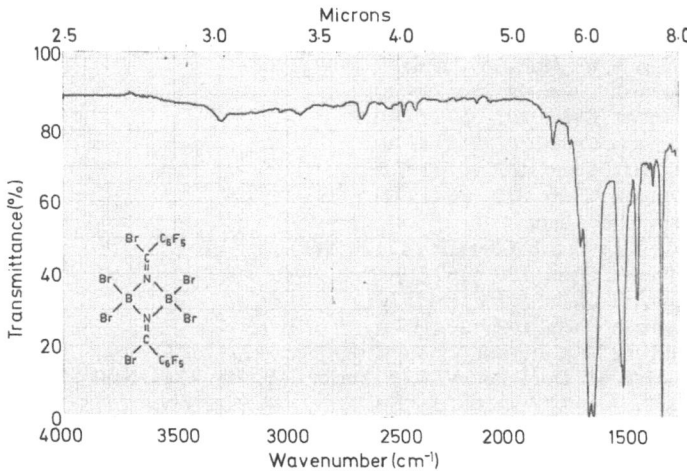

Fig. 3. I.R.-Spectrum of dimeric Pentafluorophenylbromomethleneaminodibromoborane [solution in CCl$_4$]

References

[1] Bieman, K.: "Mass Spectrometry", Organic Chemical Applications, p. 223. New York: McGraw Hill 1962.
[2] Bubnov, Ya. N., Mikhailov, B. M.: Izv. Akad. Nauk. SSSR, Ser. Khim. *1967*, 472; Bubnov, Ya. N.; Zh. Obshch. Khim. *38*, 260 (1968).
[3] Carrick, A., Glockling, F.: J. Chem. Soc. (A) *1967*, 40.
[4] Chatt, A., Richards, R. L., Newman, D. J.: J. Chem. Soc. (A) *1968*, 126.
[5] Dorokhov, V. A., Lappert, M. F.: Chem. Commun. *1968*, 250.
[6] – – J. Chem. Soc. (A) *1969*, 433.
[7] – Mikhailov, B. M.: Dokl. Akad. Nauk SSSR *187*, 1300 (1969).
[8] Emeléus, H. J., Wade, J.: J. Chem. Soc. *1960*, 2614.
[9] Gerrard, W., Lappert, M. F., Pyszora, H. W., Wallis, J. W.: J. Chem. Soc. *1960*, 2182.
[10] Hawthorne, M. F.: Tetrahedron *17*, 117 (1962).
[11] Henneberg, D.: Z. Anal. Chem. *205*, 124 (1964).
[12] Jennings, J. R., Pattison, I., Summerford, C., Wade, K., Wyatt, B. K.: Chem. Commun. *1968*, 250.
[13] – – Wade, K.: J. Chem. Soc. (A) *1969*, 565.
[14] – Wade, K.: J. Chem. Soc. (A) *1968*, 1946.
[15] Kühle, E., Anders, B., Zumach, G.: Angew. Chem. *79*, 663 (1967).
[16] Leffler, A. J.: Inorg. Chem. *3*, 145 (1964).
[17] Lloyd, J. E., Wade, K.: J. Chem. Soc. *1964*, 1649.
[18] Martius, C. A.: Liebigs Ann. Chem. *33*, 79 (1859).
[19] Meller, A.: Organometal. Chem. Rev. *2*, 1 (1967).
[20] – Topics in Current Chemistry *15*, 176 (1970).

75

21) − Marecek, H.: Monatsh. Chem. *99*, 1355 (1968).
22) − Maresch, H.: unpublished data.
23) − Maringgele, W.: Monatsh. Chem. *99*, 1909 (1968).
24) − − Monatsh. Chem. *99*, 2504 (1968).
25) − − Monatsh. Chem. *101*, 387 (1970).
26) − − Monatsh. Chem. *101*, 753 (1970).
27) − − Monatsh. Chem. *102*, 121 (1971).
28) − − Monatsh. Chem. *102*, 118 (1971).
29) − − unpublished data.
30) − Ossko, A.: Monatsh. Chem. *99*, 1217 (1968).
31) − − Monatsh. Chem. *100*, 1187 (1969).
32) − − Monatsh. Chem. *101* 1104 (1970).
33) − − Monatsh. Chem. *102*, 131 (1971).
34) − Wechsberg, M., Gutmann, V.: Monatsh. Chem. *97*, 1163 (1966).
35) Mikhailov, B. M., Dorokhov, V. A., Yakoclev, I. P.: Izv. Akad. Nauk SSSR, Ser. Khim. *1966*, 332.
36) Niedenzu, K., Dawson, J. W.: Boron-Nitrogen Compounds. In: Anorganische und allgemeine Chemie in Einzeldarstellungen, Vol. VI. Berlin-Heidelberg-New York: Springer 1965.
37) − Miller, C. D., Nahm, F. C.: Tetrahedron Letters *1970*, 2441.
38) Nöth, H., Z. Naturforsch. *16b*, 618 (1961).
39) − Vahrenkamp, H.: Chem. Ber. *99*, 1049 (1966).
40) Pattison, I., Wade, K.: J. Chem. Soc. (A) *1967*, 1098.
41) − − J. Chem. Soc. (A) *1968*, 842.
42) − − Wyatt, B. K.: J. Chem. Soc. (A) *1968*, 837.
43) Povarov, L. S., Grigos, V. I., Makhailov, B. M.: Izv. Akad. Nauk SSSR, Ser. Khim. *1963*, 2039.
44) Samuel, B., Snaith, R., Summerford, C., Wade, K.: J. Chem. Soc. (A) *1970*, 2019.
45) − Wade, K.: J. Chem. Soc. (A) *1969*, 1742.
46) Summerford, C., Wade, K.: J. Chem. Soc. (A) *1969*, 1487.
47) − − J. Chem. Soc. (A) *1970*, 2010.
48) Ardjomand, S., Lucken, E. A. C.: Helv. Chim Acta *54*, 176 (1971).

Received April 15, 1971

Atomic Absorption Spectroscopy

for the Determination of Elements in Medical Biological Samples

Professor Gary D. Christian

Department of Chemistry, University of Kentucky, Lexington, Kentucky, USA

Contents

I. Introduction

The phenomenon of atomic absorption has been known for many years. In 1802, Wollaston [1] observed the now famous absorption lines in the spectrum of the sun, called the "Fraunhofer" lines. These two lines, which occur at the same wavelengths as the sodium D-lines, were explained by Brewster [2] as being due to absorption of radiation from the sun by the layer of sodium vapor in the outer parts of its atmosphere. Kirchhoff, Bunsen, and others [3—7] demonstrated that atomic spectra could be used in emission or absorption as the basis of a new method of analysis, but the only routine use of atomic absorption for analysis until recent years was for the determination of mercury vapor contamination in laboratory atmospheres [8]. It was not until 1953, when Walsh [9] recognized the potential advantages of the flame absorption method and designed a simple apparatus for the measurement of a wide range of metals in solution, that atomic absorption came into its own as an analytical tool.

A. Principles

Walsh, in 1955, described the theoretical principles of atomic absorption spectroscopy [10]. Briefly, it can be defined as the absorption of radiant energy by ground state atomic vapor. There are several ways of obtaining atomic vapor, but aspiration of a solution into a flame is the most conventient and most widely used method. A certain fraction of the atoms produced will become thermally excited and hence will not absorb radiation from an external source. These thermally excited atoms serve as the basis of flame photometry, or flame emission spectroscopy; they can de-excite radiationally to emit radiant energy of a definite wavelength.

The ratio of excited to ground state atoms is a function of the temperature of the flame and is given by:

$$N_j/N_0 \;=\; P_j/P_0 \, \exp\,(-\Delta E/kt) = P_j/P_0 \, \exp\,(h\nu/kT) \tag{1}$$

where N_j = number of atoms in the excited state,

    N_0 = number of atoms in the ground state,

    P_j and P_0 = statistical weights of the respective states and are obtainable
                from Russell-Saunders coupling,

    ΔE = $E_j - E_0$, the energy difference between the excited and ground
                state,

    k = Boltzmann constant,

    T = the absolute temperature,

    h = the Planck's constant,

    ν = the frequency of radiant energy corresponding to the energy change,
                ΔE (this is the frequency that is absorbed or emitted).

The ratio, N_j/N_0, can therefore be calculated. For the relatively easily excited alkali metal sodium, it is 9.9×10^{-6} at $2000\,°K$ and 5.9×10^{-4} at $3000\,°K$; this latter temperature is about the highest commonly obtained with flames used for atomic absorption or emission work. Hence, only about 10^{-3} % of the sodium atoms are excited at $2000°$ and 6×10^{-2} % at $3000°$. For an element such as zinc, N_j/N_0 is 5.4×10^{-10} at $3000°$, and so only 5×10^{-8}% is excited. In spite of the small fraction excited, good sensitivities can be obtained for many elements by flame photometry if a high temperature flame is used, because the difference between zero and a small but finite number is measured. For example, seventy elements can be determined by flame photometry using the nitrous oxide-acetylene flame [11].

Atomic absorption takes advantage of the fact that most of the atoms remain in the ground state, and are capable of absorbing radiation of the appropriate wavelength corresponding to ΔE. Whereas a hot flame is preferred for flame photometry, a cooler flame is preferred for atomic absorption, except in cases where chemical interference may occur.

B. Flames

Table 1 lists the temperatures of some commonly used flames for atomic absorption. A cool flame such as argon-hydrogen-entrained air or air-coal gas is usually not preferred because of increased danger of chemical interferences (see below). The most commonly used flame is the *air-acetylene flame.*

Table 1.   *Temperatures of flames used in atomic absorption*

Flame	Maximum temperature °C
Argon-hydrogen-entrained air	1,577
Air-propane	1,725
Air-hydrogen	2,045
Air-acetylene	2,300
Nitrous oxide-acetylene	2,955
Oxygen-acetylene	3,060

It has a low burning velocity and can be used with a premix burner. The nitrous oxide-acetylene flame, also with a low burning velocity, is used for refractory elements that tend to form monoxides in the flame, preventing or decreasing production of atomic vapor. The high temperature of this flame, coupled with the highly reducing atmosphere of its red zone (due to CN and NH radicals, etc.) effectively decompose the refractory oxides to produce atomic vapor. The color-less argon-hydrogen-entrained air flame is useful for elements that absorb at very short wavelengths, less than 2000 Å (e.g., arsenic and selenium).

C. Organic Solvents

The sensitivity of atomic absorption can often be enhanced by aspirating solutions in organic solvents. The increased sensitivity is due to a number of factors, but can be attributed in large part to the lower viscosity and surface tension as compared to aqueous solutions. The flow rate is increased and smaller droplets are formed which are more efficiently vaporized. When organic solvents are aspirated, a fuel lean flame must be used in order to burn the solvent.

The best way to take advantage of the organic solvent effect without simul-taneously diluting the sample is by employing solvent extraction. By this method the element to be analyzed can actually be concentrated and a solution of the element is obtained in essentially pure organic solvent. One of the most commonly used systems involves formation of the metal chelate with ammonium 1-pyrro-lidinecarbodithioate (APDC) and then extracting this into methylisobutyl ketone (MIBK). APDC chelates of many elements form and extract into MIBK from acid solution.

Organic solvents may increase the danger of chemical interference in the flame, probably because they lower the temperature of the flame [12]. However by the technique of solvent extraction, many interferences can be eliminated.

An enhancement in sensitivity is found when hot aqueous solutions are aspi-rated, and this is attributed to essentially the same factors as for organic solvents [13].

D. Nonflame Atomizers

Nonflame cells have been used to produce atomic vapor. These include a "sputtering chamber" [14], the L'vov graphite furnace [15], and a carbon filament atom reservoir [16]. Greatly enhanced sensitivities are found for some of these. Most are not commercially available, and little work on actual applications has been described. Problems arise with biological and medical samples, due to smoke produced from the burning of the sample, which scatters the radiation from the source. However, it appears that future advances in sensitivity will be made in this direction. Several recent studies, for example, have been reported on the determination of trace elements in medicinal samples using non-flame atomizers. Anderson *et al.* [16] reported that blood samples could be analyzed using their carbon filament atom reservoir. A $1-5$ μl sample is added to a 2 mm $\times$ 4 cm carbon rod. This is heated up to about 2600 °C for 5 sec by applying up to 10 V across the filament to obtain a current of about 70 amp. The system is enclosed in a flowing argon atmosphere by a glass envelope with quartz windows. A sharp absorption peak is recorded. A detection limit of 10^{-10} g magnesium was reported and 10^{-7} g for lead. Amos *et al.* [17] have recently reported favorable results in a further study of this system. Feldman [18] has described a similar system, using a tantalum boat filament. An automatic background correction is made using a continuum source to compensate for any molecular absorption or light scattering. In this way, lead could be determined in blood using a protein-free filtrate fom a 20 μl sample.

E. Light Sources

The major requirement of the light source for atomic absorption is that it should emit the characteristic radiation (the spectrum) of the element to be determined at a half-width less than that of the absorption line. The natural absorption line width is about 10^{-4} (Å), but due to broadening factors such as Doppler and collisional broadening, the real or total width for most elements at temperatures between 2000 ° and 3000 °K is typically $0.02 - 0.1$ Å. Hence, a high resolution monochromator is not required.

The most commonly used "sharp line" source is the *hollow cathode lamp*. This consists of a hollow cathode, constructed from the element to be analyzed or from an alloy containing the element. The anode is a tungsten wire or ring. The lamp is filled under reduced pressure with an inert gas such as argon or neon. The open end of the cathode faces the window of the lamp, which is constructed of borosilicate glass or quartz; the latter window must be used if ultraviolet radiation is measured. A sufficient potential is applied across the electrodes to cause a current of from 1 to 50 mamp to flow. The inert gas is positively ionized at the anode and is accellerated at a high velocity to the cathode. Collision with the cathode causes metal atoms to "sputter" out of the cathode cup. Further collisions produce excited metal atoms which, upon de-excitation, emit the spectrum of the hollow cathode material. The spectrum of the filler gas is also emitted.

In general, only atoms in the flame that are the same as in the hollow cathode material can absorb the specific lines emitted by this material. The only requirement of the monochromator, then, is to isolate the desired line from other lines of the cathode material and the lines of the filler gas. One line of the element is usually absorbed more strongly than others (it has a higher "oscillator strength"). This often, but not necessarily, corresponds to the electronic transition from the ground state to the lowest excited state. This line is selected for maximum sensitivity measurements. For high concentrations, a line with a lower oscillator strength may be selected.

A continuous source can be used for atomic absorption, but since only the center part of the band of wavelengths passed by the slit will be absorbed (due to the sharp line nature of atomic absorption), sensitivity will be sacrificed, and the calibration curve will not be linear. This curvature is because even at high concentrations, only a portion of the radiation passing through the slit will be absorbed, and the limiting absorbance will approach a finite value rather than infinity. With a sharp line source, the entire width of the source radiation is absorbed and so the absorption follows Beer's law. A continuous source works best with the alkali metals because their absorption lines are broader than for most other elements. Specificity is not as great with a continuous source because nearby absorbing lines or molecular absorption bands will absorb part of the source.

A problem encountered with atomic absorption is that emission from the flame may fall on the detector and be registered as "negative" absorption. This can be eliminated by modulating the light source, either mechanically or electronically, and using an a.c. detector tuned to the frequency of modulation of the source. D. C. radiation, such as emission from the flame, will then not be detected. A high intensity of emission, however, may "overload" the detector, causing noise fluctuations.

F. Nonmetal Absorption

Nonmetals cannot generally be determined by direct measurement of atomic absorption in a flame because their absorption lines occur in the vacuum ultraviolet region where gases of the flame and atmosphere absorb strongly. Some can be determined by absorption of metastable lines. For example, phosphorous can be determined by the atomic absorption of a metastable line at 2135 Å. A number of indirect methods for the determination of nonmetals have been described[19].

The substance to be determined is generally reacted with a metal and the excess metal or the metal reacted with the substance is measured. Examples of biological and medical applications are given below.

G. Interferences

There are a number of interferences that can occur in atomic absorption and other flame spectroscopic methods. Anything that decreases the number of neutral atoms in the flame will decrease the absorption signal. *Chemical interference* is the most commonly encountered example of depression of the absorption signal. Here, the element of interest reacts with an anion in solution or with a gas in the flame to produce a stable compound in the flame. For example, calcium, in the presence of phosphate, will form the stable pyrophosphate molecule. Refractory elements will combine with O or OH radicals in the flame to produce stable monoxides and hydroxides. Fortunately, most of these chemical interferences can be avoided by adding an appropriate reagent or by using a hotter flame. The phosphate interferences, for example, can be eliminated by adding 1% strontium chloride or lanthanum chloride to the solution. The strontium or lanthanum preferentially combines with the phosphate to prevent its reaction with the calcium. Or, EDTA can be added to complex the calcium and prevent its combination with the phosphate.

Refractory compounds can be determined using a nitrous oxide-acetylene flame. The formation of refractory oxides with gases in flames might not be considered an interference, since it is constant under a given set of conditions; but it does decrease the sensitivity markedly so that measurement of the element may not be possible.

A second type of interference is *ionization interference.* Certain elements, particularly the alkali metals in high temperature flames, become partially ionized in the flame. This event causes a decrease in the number of neutral atoms and hence, a decrease in the sensitivity. For example, an appreciable fraction of sodium atoms will be ionized. Now if another easily ionized element such as potassium is added to the sodium solution, it will contribute free electrons to the flame and cause the equilibrium for the sodium ionization to shift toward the formation of a larger fraction of neutral atoms. This is, therefore, a positive interference. It can be overcome by adding the same amount of interfering element to the standard solution. Or, more simply, a large amount of an ionizable element such as potassium (200 to 1000 ppm) can be added to both sample and standard solutions; this will effectively suppress ionization to a small and constant value and at the same time increase the sensitivity.

Ionization interference is particularly a problem in the high temperature nitrous oxide-acetylene flame, where even elements such as manganese can be appreciably ionized. An ionization suppressant should be added for those elements having ionization potentials less than 7.5 ev. These include the alkali, alkaline earth and rare earth metals, and aluminum, chromium, gallium, indium, lead, manganese, molybdenum, niobium, scandium, tantalum, thallium, tin, titanium, vanadium, yttrium, and zirconium.

Spectral interferences are not common in atomic absorption but can occur. An element with an absorption line sufficiently close to the one of the test element that it overlaps would cause a positive interference. Fassel *et al.*[20] have discussed the problems of spectral interference. This type of interference, especially in biological samples, occurs only rarely, but the analyst should be aware of it. It is more serious if a continuous source is used. Molecular absorption is a more common spectral interference and occurs when a molecular absorption band overlaps with the atomic absorption line. For example, the CaOH species absorbs in the region of the barium 5535.5 Å line. A 1 % calcium solution gives an absorption equivalent to what is expected from about 75 ppm barium[21].

II. Biological and Medicinal Applications

Over thirty different elements have been determined in medical and biological materials by atomic absorption spectroscopy. The popularity of the technique is due to a number of factors, including sensitivity, selectivity, and ease of sample preparation. With biological fluids, often no preparation at all is required. The techniques employed usually involve simple dilution of the sample with water or with an appropriate reagent to eliminate interference. Alternatively, the element to be determined is separated by solvent extraction. Either an untreated sample, a protein free filtrate, or an ashed sample is extracted.

The method employed will depend on the concentration of the element in the sample and on the matrix interferences. Methods of sample preparation have been reviewed[22].

In the present work, emphasis is placed on summarizing recent applications of atomic absorption spectroscopy for the analysis of biological and medicinal materials. Reports prior to mid 1967 are discussed in detail elsewhere[23].

A. Blood and Urine

1. Elements Determined at Physiological Levels

Table 2 summarizes some of the metals that have been determined in blood serum and urine. The elements have been divided into those determined at their normal physiological levels, and those determined at elevated or toxicological levels in the fluids. This table lists the first group, except for the ultratrace elements to be discussed later (Table 4).

Table 2. *Elements determined in blood und urine at physiological levels*

Element	Serum, ppm [a]	Urine, mg/day [a]
Na	3200	
K	120	
Mg	36	60–120
Ca	90	96–800
Fe	0.65	0.1–0.3
Cu	1.05	0.008–0.064
Zn	1.2	0.3–0.6

[a] Physiological levels in sample.

Willis early described simple methods for the determination of the *alkali and alkaline earth metals.* Sodium and potassium are determined by direct 1 : 50 dilution of serum with water[24]. In the case of potassium, sodium is added to overcome ionization interference. Magnesium is determined by 1 : 20 dilution of serum[25] or urine[26]. With serum, 1 % strontium chloride or EDTA is added to suppress phosphate interference. Calcium is determined in serum in the same manner by adding EDTA[27]. Sodium and potassium, however, are added to the standards. For highest accuracy with serum calcium, Willis recommends preparing a protein free filtrate with trichloroacetic acid (TCA) and adding strontium chloride. Calcium is determined in urine by 1 : 20 dilution, as with magnesium[26]. Plybus et al.[28] have reported a method for determining calcium in serum utilizing a strontium internal standard. The serum is diluted 1 : 50 with 0.5 % lanthanum chloride and 10 ppm strontium and the standards have sodium and potassium added. A 0.2 – 0.3 % r.s.d. is obtained on a given day and 0.6 % r. s. d. day to day.

Several recent determinations of the alkali and alkaline earth metals in serum or urine have been reported. Barrett[29] determined potassium, sodium, and calcium in serum by diluting the samples with lanthanum chloride solution. Suttle and Field[30] used atomic absorption spectroscopy to determine potassium and magnesium in sheep plasma.

Wright and Wolff[31] analyzed *sheep serum magnesium* by diluting with 1 % hydrochloric acid containing 1 % strontium. Klein et al.[32] described a flow system for the automatic determination of magnesium in serum. Pybus[33] determined magnesium and calcium in serum and urine by diluting with strontium in perchloric acid to remove interferences from phosphorous, oxalate, and sulfate. Rodgerson and Moran [34] compared atomic absorption spectroscopy with fluorimetry for the determination of serum calcium and found better recoveries using the former method. Serum is diluted 1 : 200 with lanthanum chloride and 2-octanol. Bovine serum albumin is added to standard solutions.

Savory *et al.* [35] measured calcium and magnesium directly in protein-free filtrates of serum or urine. Baker *et al.* [36] found that both trichloroacetic acid and hydrochloric acid suppress calcium absorption and that uniform acid content is therefore required for the determination of calcium. Okuda and Sasamoto [37] determined calcium by adjusting solution conditions to 20 − 50 % methanol and 650 mg % lanthanum; serum is diluted 21-fold and urine is diluted 10−21 fold to bring the calcium concentration into the optimum range of 1 to 0.5 mg %. Osis *et al.* [38] have determined calcium and magnesium in urine, diet, and stool for metabolic studies, and Dennler and Drepper [39] have determined calcium and magnesium in the sera of sheep and calves.

Kocian [40] determined *ionized calcium* using atomic absorption spectroscopy. The ionized calcium is separated from protein bound calcium by centrifuging heparinized blood in a dialysis bag at 300 rpm for 1 h. The calcium is determined in the ultrafiltrate after diluting 25-fold. Yamaguchi and Kubushiro [41] estimated protein bound calcium in 0.5 or 1.0 ml of human serum by fractioniting the serum with gel permeation chromatography (Sephadex G−25). The eluent is 0.2M ammonium acetate at pH 6.9, and 2 ml aliquots are analyzed for calcium after adding trichloroacetic acid to precipitate proteins.

Free and protein-bound *magnesium* in normal human plasma has been determined using atomic absorption spectroscopy to measure the magnesium after the serum is ultracentrifuged and ultrafiltered [42]. A similar ultrafiltration procedure is used to determine total and filtrable magnesium and calcium in huma plasma [43] and of calcium in plasma ultrafiltrate [44], using atomic absorption spectroscopy. Briscoe and Ragan [45] described a simple empirical atomic absorption method for determining free and bound calcium in serum by means of adsorption of free ions on albumin-coated charcoal.

Hunt [46] has determined magnesium in plasma as well as in muscle and bone. Antonvewicz [47] determined magnesium in serum by precipitating proteins with trichloroacetic acid and adding strontium chloride, while Gray [48] simply diluted serum 1 : 200 with water. Other workers have recently reported the determination of magnesium in serum and plasma [49] and in plasma and urine [50], of magnesium and calcium in serum [51], and of calcium in urine [52].

Early determinations of *iron and hemoglobin* in blood were described by Herrmann *et al.* [53] and Bohmer *et al.* [54]. Zettner and co-workers [55] determinent serum iron by extracting the bathophenanthroline complex into MIBK. The serum could be diluted with water and aspirated only if the iron level was above 2 ppm. Rodgerson and Helfer [56] tried aspirating undiluted serum but obtained irreproducible results due to variations in flow rate and sample viscosity. They were able to circumvent this by integrating the signal during the aspiration of a 1 ml sample. This procedure works satisfactorily only if the iron content is less than 2.5 ppm; 6g % albumin must be added to standards to approximate the viscosity and matrix of the samples.

More recent determinations of *serum iron* have been reported by Schmidt [57], who simply diluted with lanthanum chloride solution, and by Tavenier and Hellendoorn [58], who deproteinized samples; in the latter study, iron in the protein precipitate is analyzed to correct the serum iron level. Uny *et al.* [59] determined serum iron, using ultrasonic nebulization of the sample to increase the sensitivity. Olson and Hamlin [60] have determined serum iron and total iron-binding capacity. Proteins are precipitated and iron (III) is released by heating with trichloroacetic acid.

Zettner and Mensch [61] have also determined *hemoglobin* in blood by measuring the iron in whole blood diluted 1 : 100 with water. They obtained the same results as when the sample was ashed prior to analysis. Zijlstra and Assendelft and co-workers [62,63] disagreed with this and reported differences of up to 50 % when the sample was ashed rather than diluted. The latter authors claimed serious interfrence from NaCl, KH_2PO_4 and $LaCl_3$. Zettner [64] replied that by using suitable atomizer burners, he found that these substances and a large number of other salts and acids do not effect the iron absorption.

Iron in urine has been determined by Zettner and Mansback [65] by direct aspiration, and by Devoto [66] following digestion with sulfuric acid and extraction of the iron with a 3.3 % solution of thenoyltrifluoroacetone in MIBK.

Herrmann and Lang [67] and Berman [68] first reported the determination of *serum copper* by atomic absorption spectroscopy. More recently, serum copper has been determined by Parker *et al.* [69] by dilution with water and by Dawson and co-workers [70] by dilution (1 : 20) with 0.1 *M* hydrochloric acid. Sapporo and Manabu [71] found that the addition of ethylene glycol and histidine hydrochloride to standards was essential to obtain accurate results in a simple dilution of serum. Binnerts and Achterop [72] diluted bovine plasma 1 : 6 with water to determine copper and Girard [73] diluted serum 1 : 10. Olson and Hamlin [74] diluted serum 1 : 1 with 20 % trichloroacetic acid and then heated to separate the copper from the protein. Devoto [75] prepared a protein free filtrate of serum with TCA and extracted the copper with thenoyltrifluoroacetone into MIBK. He used a similar extraction procedure for measuring copper in ashed urine samples. Schmidt [57] diluted serum samples with lanthanum chloride. Blomfield and Macmahon [76] determined total plasma copper by precipitating with 4*N* hydrochloric acid and then extracting the copper with APDC into butylacetate; trichloroacetic acid caused low results. Free plasma copper was determined in the same way, but without the hydrochloric acid treatment. O'leary and Spellacy [77] used atomic absorption spectroscopy to determine changes in total serum copper after administration of an oral contraceptive. After one month, the mean copper level rose from 1.42 ppm to 2.41 ppm.

Dawson et al. [70] aspirated urine samples directly to determine copper, but they added urinary inorganic salts to standards, which tended to suppress the copper absorption. Berge and Pflaum [78] extracted copper from urine with APDC. The copper is extracted into MIBK with the aid of an antifoaming agent, Dow Corning Antifoam B. Bojovic et al. [79] digested the urine before extracting with APDC and MIBK.

Fuwa and co-workers [80] originally determined *zinc in serum* by single 1 : 10 dilution with water. More recently, Hackley et al. [81] described a procedure in which the serum is diluted 1 : 1. Three percent dextran is added to standards to simulate the viscosity of the sample. Reinhold and co-workers [82] found that if samples are diluted 1 : 3, an ordinary capillary can be used. However, for 1 : 2 dilution, a 0.58 mm capillary is necessary for satisfactory results. Matsumoto et al. [83] diluted serum 1 : 5 with water. They observed decreased absorption in the presence of sulfuric acid and enhanced absorption in the presence of trichloroacetic acid. Haas et al. recommended diluting serum and standards 1 : 2 to 1 : 10 with water[84] or with zinc-free electrolyte model serum [85] to avoid interference by protein and serum salts. Zinc has been determined in serum and red cells by deproteinizing with trichloroacetic acid and diluting with water[86]. It has been determined in serum using the same procedures as for serum copper [73, 74]. Other workers have determined zinc in serum by direct dilution [87, 88]. McPherson and George [89] determined total copper and zinc of red cells and the free copper and zinc of plasma and dialysis fluids of patients undergoing regular hemodialysis, using atomic absorption spectroscopy. Spry and Piper [90] determined zinc in whole blood and plasma in blood cells of iron deficient rats. The zinc concentrations were raised in the iron deficient rats.

Matsumiya et al. [91] determined zinc in both serum and urine by direct dilution. Dawson and Walker [92] diluted plasma 20-fold with 0.1 N hydrochloric acid, in whole blood diluted 100-fold, and in urine diluted 10-fold. Suppression of up to 15 % of the absorbance by inorganic components was overcome by adding the appropriate amounts of those ions to the standards. Willis [93] determined zinc in urine by direct aspiration. Other workers have also determined zinc in urine [79, 94] and in blood and urine [95].

2. Elements Determined at Elevated Levels

Table 3 summarizes those elements determined at levels exceeding the physiological concentrations. Bowman [96] determined 0.3 ppm *lithium* in serum by 1 : 10 dilution, and by adding sodium and potassium to standards. The concentration of lithium in serum is about 1 to 10 ppm after administration of lithium carbonate to manic depressive patients. Zettner and coworkers [97] used a similar procedure to determine lithium while Lehmann [98] diluted with 0.1 N hydrochloric acid. Blijenberg and Leijnse [99] compared atomic absorption and atomic emission for the determination of lithium in serum and they found emission to be slightly more satisfactory. For atomic absorption, they prepared a protein-free filtrate with 96 % ethanol using 1 : 10 dilution. Trichloroacetic acid could not be

Table 3. *Elements determined in blood and urine at elevated levels*

Element	Serum, ppm [a]	Urine, ppm [a]
Li	0.3	X [b]
Sr	0.1	0.1
Mo	0.2	
Au [c]	0.1	X [b]
Hg	0.01 [c]	0.01 (0.003) [d]
B	15	
In [c]		X [b]
Tl [c]	0.005	0.1 (0.005) [e]
Pb		0.02
As		0.1 (indirect)

[a] Lowest concentration determined in sample.
[b] Determined but concentration not available.
[c] Not normally present. Others normally present but a lower levels.
[d] Nonflame absorption cell.

used because it suppressed the absorption. Woods *et al.* [100] diluted serum 1 : 1 with 0.01 % Sterox SE; they prepared standards from a synthetic serum containing sufficient glycerol to match the viscosity of the serum. The Sterox SE reduces the tendency of the serum to clog the burner and it stabilizes the solution. Tompsett [101] precipitated serum proteins with trichloroacetic acid before measuring the lithium. Zettner *et al.* [97] diluted urine 1 : 100 with water while Hansen [102] diluted urine 1 : 125 and serum 1 : 25 to determine lithium following lithium therapy.

Curnow and co-workers [103] determined *strontium* in serum and urine by coprecipitating it with calcium oxalate, dissolving the precipitate and adding lanthanum chloride to eliminate remaining traces of interferents. Puymbroeck and coworkers [104] ashed urine and then separated the strontium either by precipitation with calcium oxalate or by ion exchange chromatography. Tompsett [101] described a similar procedure using ion exchange. Descube *et al.* [105] determined 0.1 ppm strontium in biological materials.

Molybdenum (0.2 ppm) in serum has been determined by Pierce and Cholak [106] by dry ashing and taking up in water.

Gold, in the form of gold sodium thiomalate, called myochrysine, is administered in the treatment of rheumatoid arthritis. Lorber and co-workers [107] have used atomic absorption spectroscopy to monitor the gold in the serum and urine of patients by a standard additions technique. Serum is divided 1 : 1 with sodium dodecyl sulfate while urine is diluted 3 : 4 with water. Thompett [101] determined gold in 100 ml of urine by dry ashing, extracting the gold into diethyl ether from 5N hydrochloric acid, evaporating the ether, and then dissolving the residue in 1 ml of 5N hydrochloric acid.

A number of workers have described methods for the determination of *mercury* in which the mercury is first reduced to the element or collected as the sulfide on a cadmium sulfide pad. It is then volatilized into a chamber for measurement. These techniques are extremely sensitive. Thillez [108] recently described a procedure for urinary mercury in which the mercury is collected on platinum and then volatilized into an air stream. Rathje [109] treated 2 ml of urine with 5 ml of nitric acid for 3 min, diluted to 50 ml, and added stannous chloride to reduce the mercury to the element. A drop of Antifoam 60 was added and nitrogen was blown through the solution to carry the mercury vapor into a quartz end cell where it is measured. Six nanograms of mercury can be detected. Willis [93] employed more conventional methods to determine 0.04 ppm of mercury in urine by extracting it with APDC into methyl-*n*-amyl ketone. Berman [110] extracted mercury with APDC into MIBK to determine 0.01 ppm.

Boron in blood and tissue has recently been determined by Bader and Brandenberger [111] by dry ashing, and then aspirating the acidified solution into a nitrous oxide-acetylene flame. A limit of detection of 15 ppm in the solution was reported.

Torres [112] has separated *indium* from urine by ion exchange chromatography prior to determination by atomic absorption spectroscopy.

Thallium has been determined in 10 ml of ashed serum or in urine by extracting with sodium diethyldithiocarbamate into MIBK [110]. More recently, Savory and co-workers [113] described a wet digestion procedure for 50 ml of urine or 5 ml of serum in which the thallium is separated by extracting the bromide into ether, evaporating the ether and then taking up in dilute acid for aspiration. As little as 0.1 ppm is determined in urine. Curry et al. [114] determined less than 1 ng of thallium in 200 μl of urine by using the tantalum sample boat technique. The sample in the boat is dried by holding the boat 1 cm from the flame and then it is inserted into the flame where it is vaporized. A similar procedure is used for ≥3 ng of thallium in 50–100 μl of blood, except that the blood is preashed with 3 drops of nitric acid. Since the tantalum boat method is susceptible to interelement interferences, the method of standard additions is used for calibration.

Devoto [115] has described an indirect procedure for the determination of 0.1 ppm *arsenic* in urine. The arsenomolybdic acid complex is formed and extracted from 1 ml of urine at pH 2 into 10 ml of cyclohexanone. The molybdenum in the complex is then measured. Before extracting the arsenic, phosphate in the urine is separated by extracting the phosphomolybdic acid complex at pH 1 into isobutyl acetate. The direct determination of arsenic in biological material and blood and urine is best done using a nitrous oxide-acetylene flame [116]. The background absorption by this flame is low at 1937 Å, and interferences are minimized due to the high temperature of the flame.

3. Ultratrace Elements Determined

Table 4 summarizes the ultratrace elements that have been determined in serum and urine at physiological levels. These are those elements that generally occur at 0.1 ppm or less. Lead is included because it is near this level.

Table 4. *Ultratrace elements determined in serum and urine* a)

Serum:	Cr	0.03 ppm
	Mn	0.01–0.02 ppm
	Ni	0.025 ppm
	Cd	0.003–0.1 ppm
	Pbb)	0.3–0.4 ppm
Urine	Sr	<0.01–0.03 ppm
	Mn	0.001–0.01 ppm
	Ni	0.025 Mg/day (0.007–0.04)
	Cd	0.001–0.2 ppm
	Cu	0.018–0.052 mg/day; 0.006–0.03 ppm
	Bi	0.02 ppm

a) Numbers represent physiological levels in the samples.
b) Whole blood.

Feldman and co-workers [117] described a procedure for determining as little as 10 ppb of *chromium* in serum. The normal level is 30 ppb. At least 2 ml of serum are digested or dry ashed and treated with not permanganate to oxidize chromium to chromium(VI). The chromium(VI) is extracted from $3M$ HCl into 5 ml MIBK in the cold. This method has been used to measure chromium levels in studies relating this element to diabetes. Thousands of analyses have been performed. Devoto (198) dry ashed 10 ml of blood and extracted the chromium with 5 ml of 10 % tributyl phosphate in MIBK. Recently, Feldman [119] has determined

91

chromium directly in serum after 1 : 3 dilution. A background correction is made with a nonabsorbing line or a continuous source and as little as 5 ppb of chromium in the serum can be detected.

Mahoney et al. [120] were able to determine physiological levels of *manganese* in serum. A simple 1 : 1 dilution of a 2 ml sample with water is employed. The sensitivity is sufficiently high (0.005 ppm detection limit) that normal levels of manganese can be measured. A method of standard additions is used for calibration, requiring four 2 ml samples; thus, a total of 8 ml serum is required for analysis. They found a mean manganese concentration in the serum of normal fasting human subjects of 2.4±0.7 μg/100 ml. Large quantities of administered manganese did not produce a detectable change in the serum manganese concentration. Ajemian and Whitman [121] determined manganese in urine by dry ashing the urine, dissolving the residue in 10 ml of 10 % hydrochloric acid, adding 20 ml of 20 % citric acid, 5 ml of 8-hydroxyquinoline, and 3 drops of hydrogen peroxide. The pH is adjusted to 8.3 and the manganese is extracted with 20 ml of MIBK-CHCl$_3$ (1 : 1 mixture). The manganese is stripped from the organic phase with 10 ml of 10 % perchloric acid in which it is then analyzed. The detection limit is 0.02 ppm in the final solution. Ajemian and Whitman reported a normal urine manganese content of $1-10$ μg/l and that greater than 10 μg/l is indicative of increased manganese exposure. Although early reports suggested an excretion of 0.04–0.07 mg of manganese per day in the urine [122] more recent data indicate that only about 1 μg/day or approximately 0.001 ppm manganese is eliminated in the urine [123,124]; these latter data agree with those reported by Ajemian and Whitman [121].

Schaller and co-worker [125] determined *nickel* in whole blood, serum or plasma by extracting with APDC into isobutyl acetate. They reported 27 ppb nickel in blood and 21 ppb in plasma by their technique. Willis [93] described an extraction method for determining nickel in urine. The nickel in 50 ml of urine is extracted with APDC into only 1.5 ml of MIBK, and a method of standard additions is used for calibration. This method can only be used, however, to measure concentrations of 50 ppb or greater of nickel in the urine. This is just above the expected maximum normal concentration and about twice the average. Hence, this method can be used only to determine elevated nickel levels. Sunderman [126] realized this limitation and attempted other direct solvent extraction methods. These proved unsatisfactory and he resorted to wet digesting of a 50 ml sample. The nickel is isolated by extracting with dimethylglyoxime from ammoniacal solution containing citrate into chloroform. The nickel is then back extracted into 1 ml of 0.5N hydrochloric acid for aspiration. Standards are prepared in a similar manner since only 80 % of the nickel is recovered. Sunderman found a mean of 18 ppb nickel in urine by this method with a range of 4 to 31 ppb.

Lehnert et al. [127, 128] determined physiological levels of *cadmium* in serum and urine using atomic absorption spectroscopy. Ten milliliters of serum or 50 ml of dried urine are digested with nitric, sulfuric, and perchloric acids, followed by extraction of the cadmium with APDC at pH 2.5 into 2 ml of MIBK. They reported that normal adults contain $0.33 \pm 0.24\,\mu g/100$ ml of cadmium in their serum and excrete $0.98 \pm 0.36\,\mu g/$day in their urine. Other workers, however, indicate that the normal levels of cadmium may be considerably higher than this [129] ranging from 0–0.1 ppm in serum and 0–75 $\mu g/$day in the urine [130]. Klaus [131] reported an average concentration of 0.04 ppm cadmium in human serum, 0.18 ppm in human urine, and 0.03 ppm in rat urine, while Smith and Kench [132] reported a normal concentration in the urine of 0.002 to 0.022 ppm cadmium. Tipton *et al.* [133] found an average elimination of 83 and 92 μg per day in the urine of two individuals over a period of several weeks.

Berman [110] could determine as little as 0.005 ppm cadmium in serum and 0.002 ppm in urine by extracting the cadmium from the digest with lead in sodium diethyldithiocarbamate into MIBK. Torres [112] isolated cadmium from urine by ion exchange chromatography.

There is a great deal of interest in the determination of *lead*, particularly micromethods applicable to the analysis blood lead in children. Consequently, reports continue to appear on the atomic absorption determination of lead in blood and urine. Ninety percent of blood lead is found in the erythrocytes and, therefore, whole blood is analyzed rather than serum or plasma. Berman *et al.* [134] have described a procedure for determining normal lead levels in which only 250 μl of blood are taken. The blood is deproteinized with 1 ml of 10 % trichloroacetic acid and then the lead is extracted with APDC into 1 ml of MIBK, at pH 3.5. A molten lead hollow cathode lamp gives the required sensitivity for the analysis. Hessel [135] also determined physiological levels of lead in blood. Blood samples are hemolyzed with Triton X–100 to release the lead from the cells, and the lead is extracted with APDC into MIBK. Standards are made in pooled human blood. Selander *et al.* [136] used a similar procedure, but just precipitated proteins with trichloroacetic acid before extracting. Einarsson and Lindstedt [137] measured normal blood lead levels simply by precipitating proteins with trichloroacetic acid in the presence of perchloric acid and analyzing the supernatant liquid directly. Donovan and Feeley [138], on the other hand, prefer to dry ash a 5 g sample and perform a double extraction. The lead is first extracted with dithizone into chloroform at pH 9.0–9.5 in the presence of cyanide, it is back extracted into 1 % nitric acid solution, and finally is extracted at pH 2.2–2.8 with APDC into 5 ml of water-saturated MIBK. This method is used for detecting high blood lead levels in workers engaged in lead mining. Ohmori [139] studied the distribution of injected organic lead compounds in plasma and corpuscles using atomic absorption spectroscopy. Tetramethyl lead, tetraethyl lead, and mixed alkyl lead all concentrated in the plasma. Devoto [14] determined lead in 5 ml of either blood or urine, by precipitating proteins with perchloric acid and extracting the lead into 10 ml of MIBK with APDC.

Willis [93] extracted lead directly from 200 ml of urine with APDC into 1.5 ml of methyl-n-amyl ketone. He was able to determine as little as 0.02 ppm of lead. Kopito and Shwachman [141], on the other hand, co-precipitate the lead from urine with bismuth nitrate by adding ammonia. The precipitated bismuth hydroxide is dissolved in acid and this solution is aspirated. Coprecipitation of the lead is not quantitative, and so standards should be prepared in the same manner. It should be possible to employ this procedure with protein free filtrates of blood without the necessity of close pH control.

EDTA salts are used for the treatment of heavy metal poisoning. Roosels and Vanderkeel [142] were able to extract lead from urine in the presence of EDTA with dithizone by adding calcium to presumably release the lead from EDTA. In view of the fact that the formation constant of the lead-EDTA chelate is 20,000,000 times larger than that of the corresponding calcium chelate, it is doubtful that the calcium actually releases the EDTA from the lead.

Zurlo et al. [143] described a procedure similar to that used by Kopito and Schwachman. The lead in urine is separated by co-precipitation with thorium in the presence of copper(II). The precipitation is quantitative, even from the urine of subjects excreting coproporphyrins or treated with chelating agents, because the added copper liberates the chelated lead. Langford [144] wet ashed urine samples before performing a double extraction, first with diethylammonium diethyldithiocarbamate in chloroform, and second with iodide and MIBK. The yield from biological samples is 70.5 ± 11 % and for 25 μg lead, the relative error is ± 20 %. Other workers have determined lead in urine by solvent extraction [145,146] whereas Torres [112] isolated and concentrated lead from urine using ion exchange chromatography.

Montford and Cribbs [147] determined physiological levels of *strontium* in 1—5 ml of urine by co-precipitating the strontium on lanthanum carbonate and then dissolving this in 10 ml of 0.7N hydrochloric acid. Normal urine specimen showed from <0.01 to 0.03 ppm strontium; the limit of detection was 0.01 ppm. Delves [148] determined elevated strontium levels and ten other elements in the blood of normal children and children with pica, using a digestion with nitric, perchloric, and sulfuric acids, followed by a six stage solvent extraction separation procedure. Blood samples from children with pica showed abnormally high concentrations of strontium, as well as cadmium, chromium, and manganese.

The *copper* content of urine is so low that most investigators find it necessary to extract the copper into an organic solvent [78, 79]. Using a direct aspiration procedure for urine [70], Dawson et al. determined the normal daily excretion of copper to be 52 μg with a range of 26—64 μg. Sunderman and Roszel [149], on the other hand, by extracting copper from acid digested urine, found a mean urinary excretion 18.4 μg with a range of 7.5—33.8 μg; the concentration averaged 0.017 ppm with a range of 0.006—0.03 ppm.

Willis determined the physiological level of *bismuth* in urine to be about 0.02 ppm by using the same procedure he described for lead [93]. Devoto [150] dry ashed 100 ml of urine and extracted bismuth with APDC into 5 ml of MIBK.

B. Other Biological and Agricultural Materials

Atomic absorption spectroscopy has been used for the analysis of several metals in numerous other biological, medicinal, and agricultural materials. Early determinations have been summarized [23].

Those elements determined in *tissues and organs* include

Li, Na, K, Mg, Ca, Ba, Cr, Mn, Fe, Ni, Cu, Zn, Cd, Hg, B, Tl, Pb, Se and Te.

These samples are prepared by either wet or dry ashing. Many of the metals can be determined in aqueous solution, but for the more trace ones, solvent extraction procedures similar to those described above are resorted to. Similar sample preparation procedures apply to plants. The elements

Li, Na, K, Rb, Mg, Ca, Sr, Cr, Mo, Mn, Fe, Co, Ni, Cu, Ag, Zn, Pb and Se,

have been determined *in plants* while

Na, K, Rb, Mg, Ca, Sr, Cr, Mo, Mn, Fe, Co, Ni, Cu, Zn, Hg, Al, and Si

have been determined *in soils*. The metals are generally extracted from soil with ammonium acetate, ammonium chloride, acetic acid, or the like. The aqueous solution, sometimes with added reagents to minimize interferences, is often aspirated directly.

The elements determined in *feed samples* include

Na, K, Mg, Ca, Sr, Cr, Mn, Fe, Cu, Zn, and Hg,

while

Na, K, Mg, Ca, Mo, Mn, Cu, Zn and Hg

have been determined in *fertilizers*.

The Association of Official Analytical Chemists is utilizing atomic absorption spectroscopy for the determination of several trace elements. McBride [274, 275], has conducted collaborative and ruggedness studies for a number of elements in *fertilizers*. He dissolves the sample in hydrochloric acid, filters, and dilutes to volume with dilute hydrochloric acid. Standards are prepared in the same concentration of hydrochloric acid. He has recommended this procedure for adoption as official, first action for Cu, Fe, Mg, Mn, Zn, and Ca. In the case of calcium, lanthanum chloride is added to overcome phosphate interference or else a nitrous oxide-acetylene flame is used. McBride recommended further study for sodium and potassium. Cundiff and Dobbins [276] obtained very precise results for the determination of potassium in tobacco, and recommended the atomic absorption method as official first action; but results were not completely satisfactory for calcium, perhaps due to too low a flame temperature used by one collaborator. In the method, potassium and calcium are eluted from a celite column with dilute hydrochloric acid.

Rogers [265], after a collaborative study, recommended adpotion of atomic absorption as official, first action for the determination of zinc in feeds. Heckmann [226] recommended the technique for Zn, Mn and Fe in addition to Ca and Mg in feeds, but not for Cu.

Recent determinations of elements in various *biological materials* are summarized in Table 5. Many of these determinations are straight forward and require little special sample preparation. Rees and Hughes [153] observed that perchloric acid present in the digests of diet and feces samples markedly depresses the absorption of sodium and potassium when a cool air-natural gas flame is used, but the effect is reduced with an air-acetylene flame. Thalmann and Ebert [183] found interference on magnesium absorption from acid residues used for digesting plant samples, and recommended adding the acid mixture to standards, or else diluting the sample solution 1 : 20, to eliminate the interference. When Guillaumin [157] analyzed vegetable oils and fats by atomic absorption spectroscopy, the sample was dissolved in an organic solvent mixture to give an optimal fat concentration of 0.1–3 %.

Table 5. *Atomic absorption determination of metals in other biological materials*

Elements	Sample	Sample Treatment	Remarks	Ref.
Li	Biological samples			151)
K, Mg	Cerebrospinal fluid			152)
Na, K	Diet and feces	Digest	$HClO_4$ in digest decreased absorption	153)
K, Ca, Mg	Biological extracts		Add LiCl as ionization suppressent	154)
K, Na, Mg, Ca, Mn	Phospholipids	Dissolve in isopentyl acetate, add $LaCl_3$ in EtOH	Stds. same	155)

Table 5 (continued)

Elements	Sample	Sample treatment	Remarks	Ref.
K, Na, Mg, Ca, Mn, Fe, Co, Cu, Zn	Cottonseed meal	Dry ash, add $LaCl_3$		156)
K, Na, Mg, Ca, Fe, Ni, Cu	Vegetable oils and fats	Dissolve in 85 : 15 mixt. of iso-AmOAc and MeOH	Add $LaCl_3$ for Ca and Mg	157)
K, Mg, Mn	Plant tissue	Dry ash, add NH_2OH in $1N$ HCl		158)
K, Mn, Zn	Plants			159)
K, Mg, Ca	Plants	Dry ash		160)
K, Mg, Ca, Mn, Fe, Cu, Zn, Al	Plant tissue			161)
K, Na, Co, Cu, Al	Plant tissues			162)
K, Mg, Ca, Mn, Cu, Zn	Wheat	Digest w/HNO_3/$HClO_4$		163)
K	Fertilizers			164)
K	Fertilizers	Digest w/H_2SO_4		165)
K	Soil extracts	Extract with Na-tetra-phenylboron to elimi-nate interferences		166)
Na, Ca	Soils			167)
Mg	Biological materials	$0.1N$ HC1		168)
Mg, Mn, Fe Zn	Canned juices	Dilute 10– to 100– fold		169)
Mg, Zn	Mitochondria			170)
Mg, Fe, Zn	Lungs			171)
Mg, Ca	Cartilage extracts			172)
Mg, Fe, Cu, Zn, Pb	Beer and wine			173)
Mg, Ca	Rumen liquid			174)
Mg, Fe, Cu, Zn	Human brain			175)
Mg, Ca	Brain, skeletal muscle, C.S.F.	Dry ash or extract $0.5. N HNO_3$		176)
Mg	Tissue, serum, erythrocytes, kidney			177)
Mg, Ca, Fe, Cu, Zn	Rabbit tissue	Dry ash		178)

Table 5 (continued)

Elements	Sample	Sample treatment	Remarks	Ref.
Mg, Ca	Diet and stool			179)
Mg	Vomitus	Wet ash w/$HClO_4$ and HNO_3		180)
Mg, Fe, Cu, Zn	Human hair			181)
Mg	Plant material	Kjeldahl digestion	Add $SrCl_2$	182)
Mg	Plant samples	Wet ash	Add acids to stds.	183)
Mg, Mn, Fe, Co, Cu, Zn	Plants and soil extracts	Dry ash plants		184)
Mg, Ca	Subtropical forages			185)
Mg	Hay	Dry ash		186)
Mg	Food	Dry ash	Add 2500 ppm Sr to overcome Al & Si interf.	187)
Mg, Cu, Zn	Animal feeds	Dry ash		188) 189)
Mg, Ca, Mn, Fe, Cu, Zn	Animal feeds	Dry ash		190)
Mg, Ca, Mn, Fe, Cu, Zn	Fertilizers		Recommended official method	191)
Mg, Ca, Mn, Fe, Cu	Fertilizers		Adopted as official method	192)
Mg, Ca, Mn, Fe	Rice plants and soils		Add Sr or La	193)
Mg, Ca	Soil extracts	$HC_2H_3O_2$ or $NaC_2H_3O_2$ extract. Add $SrCl_2$		194)
Mg, Ca	Soil extracts	NH_4Cl, CaLactate, $NH_4C_2H_2O_3$ $-HC_2H_3O_2$, or NH_4Lactate extract. Dilute 1:100	Oxidizing flame, add Sr	195)
Mg, Ca	Soils			196)
Mg	Soil extracts		Automatic measurement	197)
Mg, Ca, Mn Cu, Zn, Al	Soils, plants	Dry ash plants		198)
Ca	Serum and urine	Dilute with $LaCl_3$	Add Na to stds.	199)

Table 5 (continued)

Elements	Sample	Sample treatment	Remarks	Ref.
Ca	Ovine fluids	Add $SrCl_2$ or $LaCl_3$	HNO_3 interferes	200)
Ca	Vegetable oils	Dissolve in MIBK and extract Ca w/2N HNO_3 −1% H_2O_2. Add $LaCl_3$	Std. addns. to calibrate	201)
Ca, Cu	Sugar beet leaves	Wet ash using Mo(VI) catalyst	Add MeOH for Cu	202)
Ca	Tooth enamel			203)
Ca	Red cells			204)
Ca	Milk			205)
Ca	Mitrochondria			206)
Ca, Sr	Algae	Treat with HNO_3 −H_2O_2		207)
Ca	Wheat internodes			208)
Ca	Grass	Wet ash		209)
Ca, Sr, Mn, Zn	Plant and soil extracts	Digest plants add $LaCl_3$		210)
Ca	Plants	Digest w/$HClO_4$−HNO_3 add $SrCl_2$		211)
Ca	Plant material	Add Mg and H_2SO_4 to samples and stds.		212)
Ca	Soil extracts	Add $LaCl_3$ and 1 % H_2SO_4		213)
Ca, Sr	Soil extracts			214)
Sr	Standard plant material			215)
Sr, Ba	Maize meal and bread flour	Dry ash, separate Sr and Ba by ion exchange		216)
Ba	Biological	Digest w/H_2SO_4, dissolve the $BaSO_4$ in $2N\ NH_3$ and 1% EDTA		101)
Cr	Tanned skins	Ash in presence of H_2SO_4		217)
Cr	Leather	Ash w/HNO_3 −H_2SO_4		218)
Cr, Mn, Fe, Cu, Zn	Plant materials			219)
Mo, Fe, Cu	Milk xanthine oxidase		Std.addns.for calibration	220)

Table 5 (continued)

Elements	Sample	Sample treatment	Remarks	Ref.
Mo	Fertilizers	Extract w/SCN$^-$ into isoamyl alc.		221)
Mn, Fe, Co, Ni, Cu, Zn	Biological materials	Solvent extraction, digest chelates		222)
Mn, Fe, Co, Ni, Cu, Zn	Biological materials		$SO_4^=$ interferes	223)
Mn, Fe, Cu, Zn	Musts & wines		Recoveries low from musts, high from wines	224)
Mn, Fe, Cu, Zn	Citrus leaf	Wet ash w/HNO_3 $-HClO_4$		225)
Mn, Fe, Cu, Zn	Feeds	Dry ash or wet ash w/HNO_3	Recommended for Zn, Mn, Fe	226)
Mn, Fe, Zn	Culture cells			227)
Mn, Cu, Zn, B	Peanut plants			228)
Mn	Fertilizers	Dissolve in HCl		229)
Mn	Soil extracts			230)
Mn, Cu, Zn	Soil solutions			231)
Mn	Soil extracts	$NH_4C_2H_3O_2$, $HC_2H_3O_2$ or Morgan's reagent		232)
Mn, Co, Ni, Cu, Zn	Soils	HCl solution or concentrate 70-fold by ion exchange		233)
Mn, Fe, Al	Soils			234)
Mn, Co, Ni, Cu, Zn	Soil extracts	$1.0N$ $HC_2H_2O_2$ required		235)
Fe, Sn	Canned fruit juice	No ashing required		236)
Fe, Cu	Wines and juices	No ashing required		237)
Fe, Cu	Cola drinks			238)
Fe, Cu, Zn	Brain tissue and hair	Chloric acid digestion		239)
Fe, Cu, Ag, Hg	Blood (carboxy-hemoglobin), liver (Cu), skin (Ag).			240)
Fe, Cu, Zn	Fatty substances, butter			241)
Fe^{+2}, Fe^{+3} Cu	Shrimp	Extract from shrimp w/HCl. Separate Fe^{+3}, Fe^{+2} by ion exchange		242)

Table 5 (continued)

Elements	Sample	Sample treatment	Remarks	Ref.
Fe, Ni, Cu	Fats and oils			243)
Fe	Leaves, seeds			244)
Fe	Soil extracts	Morgan's reagent or $NH_4C_2H_3O_2$ extract		245)
Fe, Al	Soil extracts	Oxalate extract		246)
Co	Heart tissue			247)
Co, Cu, Zn sugar	Plants			248)
Cu, Zn	Organs & tissues	Extract Zn w/HNO_3 $-HCl$ and Cu w/HNO_3 $-HClO_4$		73)
Cu	Plant materials	Soln. contains 26 ml PrOH and 0.5 ml C_6H_6 w/0.1N HNO_3		249)
Cu	Plant tissue	Dry ash, solvent extrn. w/APDC & MIBK		250)
Cu	Plants and soils	Dry ash plants; ignite soils, dissolve in $HF/HClO_4$		251)
Ag	Wine	Wet ash, extract w/dithizone		252)
Ag, Zn, Pb	Soil samples	Treat w/HNO_3 $-HCl$, take up in H_2O. Use Ta sampling boat for measurement.		253)
Zn	Human brain			254)
Zn	Rabbit uterus			255)
Zn	Enzyme			256)
Zn	Prostate			257)
Zn, Pb	Human skeleton	Dry ash		258)
Zn	Vegetable materials	Wet ash		259)
Zn	Brown seaweed			260)
Zn	Bean plant			261)
Zn	Bean plant tops			262)
Zn	Tomato plants	Dry ash		263)
Zn	Corn			264)
Zn	Foods	Wet or dry ash	Recommended as official method	265)

G. D. Christian

Table 5 (continued)

Elements	Sample	Sample treatment	Remarks	Ref.
Zn	Food, hair			91)
Zn	Fertilizers			266)
Hg	Tissues, blood	Dissolve in 20 ml $HCl+lg\ NaNO_3$, separate Hg by ion exchange, collect on CdS pad, and volatilize into tube for measurement		267)
Hg	Soil and rocks	Treat w/HNO_3 & $HClO_4$, ext. Hg w/dithizone & MIBK		268)
Si (organic)	Food and cottonseed oil	Dissolve in MBIK		269)
Si	Soils	Extract w/Tamm's reagent, oxalic acid, or ammonium oxalate		270)
Pb	Animal and plant tissue	Digest w/$HNO_3 -HClO_4 -H_2SO_4$, coppt. Pb on $SrSO_4$, convert to carbonate w/ $(NH_4)_2CO_3$, dissolve in $1N\ HNO_3$		271)
Pb	Canned tuna fish	Dry ash in presence of $Al(NO_3)_3$ & $Ca(NO_3)_2$, extract Pb w/dithizone evaporate $CHCl_3$, take up in H_2O		272)
Pb	Food	Extract Pb w/diethylammonium diethyldithiocarbamate into xylene in presence of ascorbic acid		273)

Jones and Isaac [161] compared atomic absorption spectroscopy and spark emission spectroscopy for the determination of several elements in *plant tissue.* By comparing results statistically using a t-test, no significant differences were found for calcium, manganese, iron, copper, zinc, and aluminium, but significant differences were found for potassium and magnesium at the 0.01 % level. Breck [162] made a similar comparison study for 15 elements. For copper, atomic absorption was far more sensitive, with an accuracy of ±1—2 % compared to an accuracy of 20 % obtained by optical emission.

Atomic absorption also gave better results for aluminum using a nitrous oxide-acetylene flame. In general, optical emission was more rapid.

McCracken *et al.* [164] compared atomic absorption with the tetraphenyl-boron method for determining potassium in 1190 fertilizers, and very close agreement was found between the two methods. Hoover and Reagor [165] also found good agreement between the two methods, and atomic absorption was far more rapid. They reported that the 7665 Å potassium line was more subject to interference than the less sensitive 4044 Å line. Temperli and Misteli [166] reported far better results for low concentrations of potassium in soil extracts by atomic absorption spectroscopy than by flame emission spectroscopy.

Aluminum and silicon together in plant and soil solutions depress magnesium absorption, and phosphorous, manganese, aluminum, iron, magnesium, molybdenum, copper, and silicon depress calcium absorption [193]. These interferences can be eliminated by the addition of strontium or lanthanum chloride [193–195] De Waele and Raimond [195] also found that an oxidizing flame suppresses the interference by sulfate, nitrate, chloride, and acetate in the determination of magnesium. Kabanova and Andreeva [235] reported that addition of acetic acid to soil extracts compensates for the negative influence of hydrochloric acid and of potassium, sodium, calcium, magnesium, aluminum, and iron on the absorption of copper, nickel, cobalt, zinc, and manganese. Allen and Parkinson [198] found atomic absorption to be suitable for the determination of magnesium, calcium, manganese, copper, zinc, and aluminum in plants and soils, but unsuitable for iron, cobalt, and molybdenum. David [212] reduced interferences of phosphorous, aluminum and silicon on the determination of calcium in plant material by adding magnesium and sulfuric acid to both samples and standards, and Evans and Grimshaw [213] suppressed similar interferences in soil extracts by adding lanthanum and 1 % sulfuric acid.

Schall [192] recommended that the atomic absorption determination of magnesium, calcium, manganese, iron, and copper in fertilizers should be adopted as official, first action.

A chloric acid digestion was used by Backer [239] for the preparation of tissue samples. The digest is simply diluted to determine iron, zinc, and copper. The tantalum sampling boat technique was used by Emmermann and Luecke [253] to measure lead, zinc, and silver in prepared soil solutions. White [158] treated ashed plants with hydroxylamine in 1N hydrochloric acid to reduce and dissolve oxides of manganese, prior to its determination by atomic absorption spectroscopy.

Neal [269] dissolved dimethylpolysiloxanes in MIBK, and analyzed the solution for *silicon*. Inorganic silicons commonly found in food and 1 % cottonseed oil did not interfere. Jordan [273] extracted lead from food into xylene rather than MIBK to overcome the variable solubility effect found with MIBK; reduction with ascorbic acid prevents interference from iron.

Nixon [277] compared atomic absorption spectroscopy, flame photometry, mass spectroscopy, and neutron activation analysis as methods for the determination of some 21 trace elements (<100 ppm) in hard *dental tissue* and *dental plaque*: silver, aluminum, arsenic, gold, barium, chromium, copper, fluoride, iron, lithium, manganese, molybdenum, nickel, lead, rubidium, antimony, selenium, tin, strontium, vanadium, and zinc. Brunelle [278] also described procedures for the determination of about 20 elements in soil using a combination of atomic absorption spectroscopy and neutron activation analysis.

The principles and applications of atomic absorption spectroscopy to clinical and biological analysis have been reviewed by several authors [279–286] and automation in the analysis has been reviewed [287].

C. Determination of Nonmetals in Medical and Biological Materials

An interesting application of atomic absorption spectroscopy is the indirect determination of nonmetals. Christian and Feldman [19] have described the various indirect methods that can be used. Methods have been described for the determination of several nonmetals in biological samples.

Helmerson [288], and more recently Bartels [289], determined *chloride* in serum by adding excess silver and then measuring the excess silver ion in the filtrate. Ezell [290] used a similar procedure to determine chloride in plant liquors and Gutsche et al. [291] determined chlorine in milk by measuring the excess silver by flame photometry.

Several investigators have described the indirect determination of *orthophosphate* by extraction of the phosphomolybdic acid complex and the measuring the molybdenum extracted. Zaugg and Knox [292] first applied this technique to the determination of phosphate in urine. A protein-free filtrate was formed and the complex was extracted into 2-octanol. More recently, Devoto [293] determined 0 to 25 μg of phosphate in 50 ml of urine by extracting the complex from acidified urine into isobutyl acetate.

Roe and co-workers [294] analyzed for *sulfate* in urine by precipitating $BaSO_4$, dissolving the precipitate in EDTA and then measuring the equivalent amount of barium. Galindo et al. [295] used a similar procedure for determining sulfur in soil extracts. Organic matter is destroyed and sulfur is oxidized to sulfate by 30 % hydrogen peroxide solution, and then barium chloride is added. $(NH_4)_2$ EDTA is used to dissolve the precipitate. Gersonde [296] determined *sulfur in proteins* by oxidizing it to sulfate in a Schöniger combustion flask and then precipitating $SrSO_4$. The strontium is measured by flame emission. Suzuki *et al.* [297]) determined protein thiol groups by measuring the mercury bound to the protein after reaction with mercuric chloride or *p*-chloromercuribenzoate. The excess unreacted reagent is separated by gel filtration, and a limit of detection of 0.1 ppm mercury is reported; this is 30-fold more sensitive than conventional spectrophotometric methods. Fuwa and Vallee [298]) determined sulfur directly in amino acids by molecular flame absorption spectrometry. The absorption of the sulfur peak at 207 nm in an air-hydrogen flame is measured with a detection limit of 10 ppm sulfur.

Bond and Willis [299]) have determined *ammonia* in mitrochondria samples after separation by distillation. The enhancement of zirconium absorption by the ammonia is measured. This is more rapid than colorimetric procedures. Hall *et al.* [300]) described an indirect method for the determination of *urinary α-amino nitrogen.* Copper is solubilized from insoluble copper phosphate by complexing with α-amino groups at slightly alkaline pH. The remaining copper phosphate is removed by filtration and the filtrate is diluted 1 : 10 or 1 : 20 with 0.1*N* hydrochloric acid to measure the dissolved copper by atomic absorption. Standards are prepared using alanine.

Potter and co-workers [301]) determined *reducing sugars* in plant materials by reducing copper(II) in alkaline solution to insoluble cuprous oxide and then measuring the excess copper in the filtrate. Mitschell [248]) conducted further studies on this method. Christian and Feldman [19]) have described general procedures for the indirect determination of glucose and of protein. It is anticipated that in the future, we will see many more applications of atomic absorption spectroscopy to the indirect determination of nonmetals.

III. Conclusion

Atomic absorption spectroscopy has proved to be an extremely effective and quite simple tool for the analysis of metals and several nonmetals in medical and biological materials. At present, only about one half of those trace elements that occur in the body have been analyzed at the physiological level by this technique, but it will probably be only a matter of time before methods are described for these elements. Other elements of interest include vanadium, molybdenum, cobalt, aluminum, arsenic, bismuth, and selenium. Also, the use of high temperature flames for flame emission spectroscopy [11, 302, 303]) will undoubtedly find use for the analysis of more trace elements, using similar techniques described here. The application of atomic fluorescence spectroscopy should prove extremely interesting, particularly for such elements as cadmium. Nonflame atom reservoirs show great promise for ultratrace analysis, requiring only small samples with little or no sample preparation.

G. D. Christian

IV. References

1) Wollaston, H. H.: Phil. Trans. Roy. Soc. London, Ser. A *92*, 365 (1802).
2) Brewster, D.: Report of the 2nd Meet., British Assoc. 320 (1832).
3) Kirchoff, G.: Pogg. Ann. *109*, 275 (1860).
4) –, Bunsen, R.: Pogg. Ann. *110*, 161 (1860).
5) – – Pogg. Ann. *113*, 337 (1861).
6) – – Phil. Mag. *22*, 329 (1861).
7) Liveing, G., DeMar, J.: Collected Papers. London: Cambridge University Press 1915.
8) Woodson, T. T.: Rev. Sci. Instr. *10*, 308 (1939).
9) Walsh, A.: Handbook of Third Exhibition of Institute of Physics, p. 42. Melbourne: Victorian Division 1954.
10) – Spectrochim. Acta 7, 108 (1955).
11) Christian, G. D., Feldman, F. J.: 21st Pittsburgh Conference on Analytical Chemistry and Applied Spectroscopy, Cleveland, March 1–6, 1970.
12) Pickett, E. E., Koirtyohann, S. R.: Anal. Chem. *41* (14), 28 A (1969).
13) Feldman, F. J., Christian, G. D.: Can. Spectry. *13*, 139 (1968).
14) Gatehouse, B. M., Walsh, A.: Spectrochim. Acta *16*, 602 (1960).
15) L'vov, B. V.: Spectrochim. Acta *17*, 761 (1961).
16) Anderson, R. G., Maines, I. S., West, T. S.: International Atomic Absorption Spectroscopy Conference, Sheffield, July 14–18, 1969.
17) Amos, M. D., Bennett, P. A., Lung, P. W. Y., Thomas, G. P.: 21st Pittsburgh Conference on Analytical Chemistry and Applied Spectroscopy, Cleveland, March 1–6, 1970.
18) Feldman, F. J.: 21st Pittsburgh Conference on Analytical Chemistry and Applied Spectroscopy, Cleveland, March 1–6, 1970.
19) Christian, G. D., Feldman, F. J.: Anal. Chim. Acta *40*, 173 (1968).
20) Fassel, V. A., Rasmuson, J. D., Cowley, T. G.: Spectrochim. Acta *23B*, 579 (1968).
21) Koirtyohann, S. R., Pickett, E. E.: Anal. Chem. *38*, 585 (1966).
22) Christian, G. D.: Anal. Chem. *41* (1), 24A (1969).
23) – Feldman, F. J.: Atomic Absorption Spectroscopy. Applications in Agriculture, Biology and Medicine. New York: Wiley-Interscience 1970.
24) Willis, J. B.: Spectrochim. Acta *16*, 551 (1960).
25) – Spectrochim. Acta *16*, 273 (1960).
26) – Anal. Chem. *33*, 556 (1961).
27) – Spectrochim. Acta *16*, 259 (1960).
28) Plybus, J., Feldman, F. J., Bowers, G. N.: Clin. Chem., *in press*
29) Barrett, D. F.: Anal. Advan. *1*, 14 (1968).
30) Suttle, N. F., Field, A. C.: Brit. J. Nutr. *23*, 81 (1969).
31) Wright, D., Wolff, J.: N. Z. J. Agr. Res. *12*, 287 (1969).
32) Klein, B., Kaufman, J. H., Oklander, M.: Clin. Chem. *13*, 788 (1967).
33) Pybus, J.: Clin. Chim. Acta *23*, 309 (1969).
34) Rodgerson, D. O., Moran, I. K.: Clin. Chem. *14*, 1206 (1968).
35) Savory, J., Wiggins, J. W., Heintges, M. G.: Am. J. Clin. Pathol. *51*, 720 (1969).
36) Baker, R. A., Hartshorne, D.J., Wilshire, A. G.: At. Absorption Newsletter *8*, 44 (1969).

37) Okuda, M., Sasomoto, H.: Rinsho Byori *17*, 380 (1969).
38) Osis, D., Royston, K., Samackson, J., Spencer, H.: Develop. Apppl. Spectry. *7A*, 227 (1968).
39) Dennler, H. J., Drepper, K.: Tieraertzl. Wochenschr. *82*, 58 (1969).
40) Kocian, J.: Cesk. Gastroenterol. Vyziva *23*, 192 (1969).
41) Yamaguchi, R., Kubushiro, J.: Bunseki Kagaku *18*, 734 (1969).
42) Nielson, S.P.: Scand. J. Clin. Lab. Invest. *23*, 219 (1969).
43) Osmum, J. R.: Am. J. Med. Technol. *33*, 448 (1967).
44) Robertson, W. G.: Clin. Chim. Acta *24*, 149 (1969).
45) Briscoe, A. M., Ragan, C.: J. Lab. Clin. Med. *69*, 351 (1967).
46) Hunt, B. J.: Clin. Chem. *15*, 979 (1969).
47) Antonvewicz, A.: Roczniki Nauk Rolniczych Sec. B. *90*, 419 (1968).
48) Gray, J. M.: Flame Notes *4*, 2 (1969).
49) Maurat, J. P., Rousselet, F.: Problems Actuels De Biochimie Appliques *2-e, 1968* 153–217.
50) Steele, T. H., Wen, S., Evenson, M. A., Rieselback, R. E.: J. Lab. Clin. Med. *71*, 455 (1968).
51) Akgin, S., Rudman, D., Wertheim, A. R.: Endocrinology *84*, 347 (1969).
52) King, J. S., Jr., Buchanan, R.: Clin. Chem. *15*, 31 (1969).
53) Herrmann, R., Lang, W., Stamm, D.: Blut *11*, 135 (1965).
54) Bohmer, M., Aver, E. A., Bartels, H.: Aertzl. Lab. *13*, 258 (1967).
55) Zettner, A., Sylvia, L. C., Capacho-Delgado, L.: Am. J. Clin. Pathol. *45*, 533 (1966).
56) Rodgerson, D. O., Helfer, E.: Clin. Chem. *12*, 388 (1966); Am. J. Clin. Pathol. *46*, 63 (1966).
57) Schmidt, W.: Fresenius' Z. Anal. Chem. *243*, 198 (1968).
58) Tavenier, P., Hellendoorn, H. B. A.: Clin. Chim. Acta *23*, 47 (1969).
59) Uny, G., Brule, M., Spitz, J.: Ann. Biol. Clin. (Paris) *27*, 387 (1969).
60) Olson, A. D., Hamlin, W. B.: Clin. Chem. *15*, 438 (1969).
61) Zettner, A., Mensch, A. H.: Am. J. Clin. Pathol. *48*, 225 (1967).
62) Van Assendelft, O. W., Zijlstra, W. G., Buursma, A., Van Kampen, E. J., Hoeck, W.: Clin. Chim. Acta *22*, 281 (1968).
63) Zijlstra, W. G., Van Assendelft, O. W., Van Kampen, E. J., Hoeck, W.: Am J. Clin. Pathol. *50*, 513 (1968).
64) Zettner, A.: Am. J. Clin. Pathol. *50*, 514 (1968).
65) – Mansback, L. M.: Am. J. Clin. Pathol. *44*, 517 (1965).
66) Devoto, G.: Russ. Med. Sarda. *71*, 357 (1968).
67) Herrmann, R., Lang, W.: Z. Klin. Chem. *1*, 182 (1963).
68) Berman, E.: Clin. Chem. *9*, 459 (1963).
69) Parker, M. M., Humoller, F. L., Mahler, D. J.: Clin. Chem. *13*, 40 (1967).
70) Dawson, J. B., Ellis, D. J., Newton-John, H.: Clin. Chim. Acta *21*, 33 (1968).
71) Sapporo, T., Manabu, N.: Clin. Chim. Acta *24*, 299 (1969).
72) Binnerts, W. T., Achterop, Th.: Tijdschr. Diergeneesk. *92*, 639 (1967).
73) Girard, M. L.: Clin. Chim. Acta *20*, 243 (1968).
74) Olson, A. D., Hamlin, W. B.: At. Absorption Newsletter *1*, 69 (1968).
75) Devoto, G.: Boll. Soc. Ital. Biol. Sper. *44*, 1249 (1968).
76) Blomfield, J., Macmahon, R. A.: J. Clin. Pathol. *22*, 136 (1969).
77) O'Leary, J. A., Spellacy, W. N.: Science *162*, 682 (1968).
78) Berge, D. G., Pflaum, R. T.: Am. J. Med. Technol. *34*, 725 (1968).
79) Bojovic, V., Stojadinovic, Lj., Djuric, D.: Med. Lavoro *59*, 357 (1968).

80) Fuwa, K., Publido, P., McKay, R., Vallee, B. L.: Anal. Chem. *36*, 2407 (1964).
81) Hackley, B. M., Smith, J. C., Halstead, J. A.: Clin. Chem. *14*, 1 (1968).
82) Reinhold, J. G., Pascoe, E., Kfoury, G. A.: Anal. Biochem. *25*, 557 (1968).
83) Matsumoto, H., Tsunmatsu, K., Shiraishi, T.: Bunseki Kagaku *17*, 703 (1968).
84) Haas, Th., Lehnert, G., Schaller, K. H.: Z. Klin. Chem. Klin. Biochem. *5*, 218 (1967).
85) – – – Beckman Rep. *1967* (2–3) 3.
86) Oiwa, K., Kimura, T., Mikino, H., Okuda, M.: Bunseki Kagaku *17*, 810 (1968).
87) Roth, M. E., Ramirez-Munoz, J.: Flame Notes *4*, 25 (1969).
88) Muranaka, H., Kanou, T.: Rinsho Byori *17*, 559 (1969).
89) McPherson, J., George, C. R. P.: Brit. Med. J. *2*, 141 (1969).
90) Spry, C. J. F., Piper, K. G.: Brit. J. Nutr. *23*, 91 (1969).
91) Matsumiya, K., Yoshinaga, T., Omori, K.: Rinsho Byori *16*, 103 (1968).
92) Dawson, J. B.,Walker, B. E.: Clin. Chim. Acta *26*, 465 (1969).
93) Willis, J. B.: Anal. Chem. *34*, 614 (1967).
94) Allan, R. E., Pierce, G. O., Yeager, D.: Am. Ind. Hyg. Assoc. J. *29*, 469 (1968).
95) Hasegawa, N., Hirai, A., Kashiwagi, T.: Ann. Rep. Res. Inst. Environ. Med., Nagoya Univ. *16*, 1 (1968).
96) Bowman, J. A.: Anal. Chim. Acta *37*, 465 (1967).
97) Zettner, A., Rafferty, K., Jarecki, H. J.: At. Absorption Newsletter *7*, 32 (1968).
98) Lehman, V.: Clin. Chim. Acta *20*, 523 (1968).
99) Blijenberg, B. G., Leijnse, B.: Clin. Chim. Acta *19*, 97 (1968).
100) Woods, A. E., Crowder, R. D., Coates, J. T., Wittrig, J. J.: At. Absorption Newsletter *7*, 85 (1968).
101) Tompsett, L. L.: Proc. Assoc. Clin. Biochem. *5*, 125 (1968).
102) Hansen, J. L.: Am. J. Med. Technol. *34*, 625 (1968).
103) Curnow, D. C., Gutteridge, D. G., Horgan, E. D.: At. Absorption Newsletter *7*, 45 (1968).
104) Van Puymbroeck, S., Jacquemin, R., Colard, J., Kirchmann, R., Van der Borght, O.: Nucl. Activ. Tech. Life Sci. Proc. Symp. Amsterdam *1967*, 267.
105) Descube, J., Roques, N., Rousselet, F., Girard, A. M. L.: Anal. Biol. Clin. *25*, 1011 (1967).
106) Pierce, J. O., Cholak, J.: Arch. Environ. Health *13*, 208 (1966).
107) Lorber, A., Cohen, R. L., Chang, E. E., Anderson H. E.: Arthritis Rheum. *11*, 170 (1968).
108) Thillez, G.: Chim. Anal (Paris) *50*, 226 (1968).
109) Rathje, A. O.: Am. Ind. Hyg. Assoc. J. *30*, 126 (1969).
110) Berman, E.: At. Absorption Newsletter *6*, 57 (1967).
111) Bader, H., Brandenberger, H.: At. Absorption Newsletter *7*, 1 (1968).
112) Torres, F.: U. S. At. Energy Comm. *1968* SC-TM-68-4.
113) Savory, J., Roszel, N. O., Mushak, P., Sunderman, F. W., Jr.: Am. J. Clin. Pathol. *50*, 505 (1968).
114) Curry, A. S., Read, J. F., Knott, A. R.: Analyst *94*, 744 (1969).
115) Devoto, G.: Boll Soc. Ital. Biol. Sper. *44*, 425 (1968).
116) Christian, G. D., Park, P. J.: unpublished data.
117) Feldman, F. J., Knoblock, E. C., Purdy, W. C.: Anal. Chim. Acta *38*, 489 (1967).
118) Devoto, G.: Boll. Soc. Ital. Biol. Sper. *44*, 1251 (1968).
119) Feldman, F. J.: private communication, 1969.
120) Mahoney, J. P., Sargent, M. G., Small, W. J.: Clin. Chem. *15*, 312 (1969).
121) Ajemian, R. S., Whitman, N. E.: Am. Ind. Hyg. Assoc. J. *30*, 52 (1969).
122) Kent, N. L., Mc Cance, R. A.: Biochem. J. *35*, 877 (1941).
123) Kanabrocki, E. L., Case, L. F., Fields, T., Graham, L., Oester, Y. T., Kaplan, E.: Developments in Applied Spectroscopy, Vol. V. (Pearson, L. R., Grove, E. L., eds.) New York: Plenum 1966.

124) Hamoguchi, T., Horiuchi, K., Tanaka, N.: Bunseki Kagaku *15*, 1264 (1966).
125) Schaller, K. H., Kuehner, A., Lehnert, G.: Blut *17*, 155 (1968).
126) Sunderman, F. W., Jr.: Am. J. Clin. Pathol. *44*, 182 (1965).
127) Lehnert, G., Schaller, K. H., Haas, Th.: Z. Klin. Chem. Klin. Biochem. *6*, 174 (1968).
128) – Klavis, G., Schaller, K. H., Haas, Th.: Brit. J. Ind. Med. *26*, 156 (1969).
129) Pinkerton, C.: private communication, 1969.
130) Gallob-Hausmann, G.: private communication, 1969.
131) Klaus, R.: Z. Klin. Chem. *4*, 299 (1966).
132) Smith, J. C., Kench, J. E.: Brit. J. Ind. Med. *14*, 270 (1957).
133) Tipton, I. H., Stewart, P. L., Dickson, J.: Health Phys. *16*, 455 (1969).
134) Berman, E., Valavanis, V., Dubin, A.: Clin. Chem. *14*, 239 (1968).
135) Hessel, D. W.: At. Absorption Newsletter *1*, 55 (1968).
136) Selander, S., Cramer, K., Borjesson, B., Mandorf, G.: Brit. J. Ind. Med. *25*, 209 (1968).
137) Einarsson, O., Lindstedt, G.: Scand. J. Clin. Lab. Invest. *23*, 367 (1969).
138) Donovan, P. P., Feeley, D. T.: Analyst *94*, 879 (1969).
139) Ohmori, K.: Yokohama Igaku *20*, 210 (1969).
140) Devoto, G.: Boll. Soc. Ital. Biol. Sper. *44*, 421 (1968).
141) Kopito, L., Shwachman, H.: J. Lab. Clin. Med. *70*, 326 (1967).
142) Roosels, D., Vanderkeel, J. V.: At. Absorption Newsletter *7*, 9 (1968).
143) Zurlo, N., Griffini, A. M., Colombo, G.: Anal. Chim. Acta *47*, 203 (1969).
144) Langford, J. C.: Anal. Chem. *41*, 1716 (1969).
145) Selander, S., Cramer, K., Borjesson, B., Mandorf, G.: Brit. J. Ind. Med. *25*, 139 (1968).
146) Taira, Y.: Nippon Eiseikensa Gishikoi Zasshi *18*, 389 (1969).
147) Montford, B., Cribbs, S. L.: At. Absorption Newsletter *8*, 77 (1969).
148) Delves, H. T.: Biochem. J. *112*, 34 P (1969).
149) Sunderman, F. W., Jr., Roszel, N. O.: Am. J. Clin. Pathol. *48*, 286 (1967).
150) Devoto, G.: Boll. Soc. Ital. Biol. Sper. *44*, 1253 (1968).
151) Little, B. R., Platman, S. R., Fieve, R. R.: Clin. Chem. *14*, 1211 (1968).
152) Bito, L. Z.: Science *165*, 81 (1969).
153) Rees, E. D., Hughes, R. D.: Clin. Chim. Acta *21*, 515 (1968).
154) Varghese, F. T. N., Lipton, A., Huxham, G. J.: Lab. Pract. *18*, 419 (1969).
155) Montford, B., Scribbs, S. C.: Talanta *16*, 1079 (1969).
156) Turner, W. W., Jr.: J. Am. Oil Chem. Soc. *44*, 129 (1967).
157) Guillaumin, R.: Rev. Franc. Corps Gras *16*, 497 (1969).
158) White, R. P.: Soil Sci. Soc. Am. Proc. *33*, 478 (1969).
159) Fuehring, H. D.: Agron. J. *61*, 594 (1969).
160) Ward, G. M., Miller, M. J.: Can. J. Plant Sci. *49*, 53 (1969).
161) Jones, J. B., Jr., Isaac, R. A.: Agron. J. *61*, 393 (1969).
162) Breck, F.: J. Assoc. Offic. Anal. Chem. *51*, 132 (1968).
163) Draycott, A. P., Durrant, M. J.: J. Agri. Sci. *72*, 319 (1969).
164) McCracken, M. L., Webb, H. M., Hammar, H. E., Loadholt, C. B.: J. Assoc. Offic. Anal. Chem. *50*, 5 (1967).
165) Hoover, W. L., Reagor, J. C.: J. Assoc. Offic. Anal. Chem. *51*, 211 (1968).
166) Temperli, A. T., Misteli, H.: Anal. Biochem. *27*, 361 (1969).
167) O'Conner, G. A., Kemper, W. D.: Soil Sci. Soc. Proc. *33*, 464 (1969).
168) Allen, J. E.: Analyst *83*, 466 (1968).
169) Ramirez-Munoz, J., Roth, M. E.: Flame Notes *4*, 10 (1969).
170) Brimley, G. P., Knight, V. A.: Biochemistry *6*, 3892 (1967).

G. D. Christian

171) Crable, J. V., Keenan, R. G., Kinsu, R. E., Smallwood, A. W., Mauer, P. A.: Am. Ind. Hyg. Assoc. J. *29*, 106 (1968).
172) Alcock, N. W., Shils, M. E.: Biochemistry J. *112*, 505 (1969).
173) Weiner, J. P., Taylor, L.: J. Inst. Brewing *75*, 195 (1969).
174) Chauvaux, G.: Ann. Med. Vet. *112*, 319 (1968).
175) Harrison. W. W., Martin, G., Brown, M. D.: Clin. Chim. Acta *21*, 55 (1968).
176) Kleeman, C. R., Bagdoyan, H., Berberian, A.: J. Lab. Clin. Med. *71*, 884 (1968).
177) Lim, P., Jacob, E., Dong, S., Khoo, O. T.: J. Clin. Pathol. *22*, 417 (1969).
178) Kitaguchi, K., Nishimoto, S.: Rinsho Byori *19*, 495 (1969).
179) Husdan, H., Rapoport, A.: Clin. Chem. *15*, 669 (1969).
180) Payne, C. E., Combs, H. F.: Appl. Spectry. *22*, 786 (1968).
181) Harrison, W. W., Yurachek, J. P., Benson, C. A.: Clin. Chim. Acta *23*, 83 (1969).
182) Collier, R. E.: Hilger J. *10*, 52 (1967).
183) Thalman, E., Ebert, K.: Albrecht-Thaer-Arch. *11*, 719 (1967).
184) Štupar, J., Furlan, J., Glazer, I.: Landwirtsch. Forsch. *20*, 12 (1967).
185) Lee, J., Campbell, C. M.: J. Dairy Sci. *52*, 121 (1969).
186) Antonvewicz, A.: Roszniki Nauk Rolniczych Ser. B. *90*, 419 (1968).
187) Osada, H., Goto, I.: Eiyo To Shokuryo *20*, 349 (1968).
188) Roach, A. G., Sanderson, P., Williams, D. R.: Analyst *93*, 42 (1968).
189) Williams, D. R.: Spectrovision *19*, 8 (1968).
190) Perrin, C. H., Ferguson, P. A.: J. Assoc. Offic. Anal. Chem. *51*, 654 (1968).
191) McBride, C. H.: J. Assoc. Offic. Anal. Chem. *51*, 847 (1968).
192) Schall, E. D.: J. Assoc. Offic. Anal. Chem. *52*, 217 (1969).
193) Kosuge, N., Ito, J., Ito, H.: Hokuriku Nogyo Shikensho Hokoku *9*, 47 (1968).
194) MacPhee, W. S. G., Ball, D. F.: J. Sci. Food Agr. *18*, 376 (1967).
195) DeWaele, M., Raimond, Y.: Bull. Rech. Agron. Gembloux *2*, 613 (1967).
196) Weed, S. B., Lenoard, R. A.: Soil Sci. Soc. Am. Proc. *32*, 335 (1968).
197) Griffin, G. F.: Soil Sci. Soc. Am. Proc. *32*, 803 (1968).
198) Allen, S., Parkinson, J.: Spectrovision *22*, 2 (1969).
199) Kocian, J., Rubeska, I.: Cesk. Gastroenterol. Vyziva *22*, 188 (1968).
200) Nel, J. W., Moir, R. J.: S. African J. Agr. Sci. *11*, 153 (1968).
201) Foss, R. A., Houston, D. M.: At. Absorption Newsletter *8*, 82 (1969).
202) Harrison, M., Andre, C.: Appl. Spectry *23*, 354 (1969).
203) Brudevold, F., McCann, H. G., Groen, P.: Arch. Oral Biol. *13*, 877 (1968).
204) Schatzmann, H. J., Uincenzi, F. F.: J. Physiol. *201*, 369 (1969).
205) Southgate, D. A. T., Widdowson, E. M., Smits, B. J., Cooke, W. T., Walker, C. H. M., Mathers, N. P.: Lancet *1*, 487 (1969).
206) Patriarca, P., Carafoli, E.: Experientia *25*, 598 (1969).
207) Fuller, W. G., Hardcastle, J. E.: Soil Sci. Soc. Am. Proc. *31*, 772 (1967).
208) Parish, W., Miller, F. L.: Australian J. Biol. Sci. *22*, 77 (1969).
209) Elzam, O. W., Epstein, E.: Agrochimica *13*, 187 (1969).
210) Halstead, E. W., Barber, S. A., Warncke, D. D., Bole, J. B.: Soil Sci. Soc. Am. Proc. *32*, 69 (1968).
211) Loneragan, J. F., Snowball, K.: Australian J. Agri. Res. *20*, 465 (1969).
212) David, D. J.: Analyst *84*, 536 (1969).
213) Evans, C. C., Grimshaw, H. M.: Talanta *15*, 413 (1968).
214) Juo, A. S. R., Barber, S. A.: Soil Sci. Soc. Proc. *33*, 360 (1969).
215) David, D. J.: Analyst *94*, 884 (1969).
216) Strelow, F., Norval, E.: J. S. African Chem. Inst. *20*, 25 (1967).
217) Montagut, M., Obiels, J., Genis, I.: Afinidad *25*, 443 (1968).

218) Montagut, M., Obiels, J., Genis, I.: AQEIC (Asoc. Quim. Espan. Ind. Cuero) Bol. Tec. *19*, 309 (1968).
219) Premi, P. R., Cornfield, A. H.: Spectrovision *29*, 15 (1968).
220) Roussos, G. G., Morrow, B. H.: Apl. Spectry. *22*, 769 (1968).
221) Hoover, W. L., Duren, S. C.: J. Assoc. Offic. Anal. Chem. *50*, 1269 (1969).
222) Novikova, N. N., Tereschchenko, A. P., Byr'ko, V. M.: Mater. Rab. Soveshch, Vop. Krugovorota Veshchesto Zamknutoi Sist. Osn. Zhiznedeyatel. Nizshikh Organizmov 5th, *1967*, 57-9.
223) Zhurenko, V. N., Tereschchenko, A. P., Tokareva, L. N., Sheveleva, R. L.,: Mater. Rab. Soveshch. Vop. Krugovorota Veshchesto Zamknutoi Sist. Osn. Zhiznedeyatel. Nizshikh Organizmov 5th, *1967*, 61-4.
224) Amati, A., Rastelli, R., Minguzzi, A.: Ind. Agr. (Florence) *6*, 630 (1968).
225) Steyn, W. J. A., Krueger, H.: Tydskr. Natuurwetensk. *9*, 22 (1969).
226) Heckman, M.: J. Assoc. Offic. Anal. Chem. *51*, 776 (1968).
227) Hayward, J.: J. Marine Biol. Assoc. U. K. *49*, 439 (1969).
228) Martens, D. C., Hallock, D. L., Alexander, M. W.: Agronomy J. *61*, 85 (1969).
229) Giordano, P. M., Mortvedt, J. J.: Soil Sci. Soc. Am. Proc. *33*, 460 (1969).
230) Browman, M. G., Chesters, G., Ronke, H. B.: J. Agri. Sci. *72*, 335 (1969).
231) Grimme, H.: Z. Pflanzenernaehr. Bodenk. *121*, 58 (1968).
232) Nadirshaw, M., Cornfield, A. H.: Analyst *93*, 475 (1968).
233) Andreeva, T. B., Kabanova, M. A.: Khim. Sel. Khoz. *6*, 869 (1968).
234) Road, A. T., Protzl, R., Thomas, R. L.: Soil Sci. *49*, 89 (1969).
235) Kabanova, M. A., Andreeva, T. B.: Khim. Sel. Khoz. *7*, 552 (1969).
236) Price, W. J., Roos, J. T. H.: J. Sci. Food Agr. *20*, 437 (1969).
237) Lipis, B. V., Pchelintsev, A. M., Spektor, L. A.: Sadovod. Vinograd. Vinodel. Mold. *5*, 38 (1968).
238) Pascual, P. S.: Sci. Bull. Sci. Found. Phillipp. *13*, 17 (1968).
239) Backer, E. T.: Clin. Chim. Acta *24*, 233 (1969).
240) Brandenberger, H.: Ann. Biol. Clin. (Paris) *25*, 1053 (1967).
241) Rostelli, R., Amati, A.: Riv. Ital. Sostanze Grasse *46*, 62 (1962).
242) Thompson, M. H., Farragut, R. N.: Fishery Ind. Res. *5*, 2 (1965).
243) Persmark, U.: Metal Catal. Lipid Oxid., SIK (Sv. Inst. Konserveringsforsk.) Symp., Pap. Discuss. *1967*, 43-7.
244) Matocha, J. E., Thomas, G. W.: Agron. J. *61*, 425 (1969).
245) Khan, F. R., Cornfield, A. H.: Plant Soil *29*, 189 (1968).
246) Blakemore, L. C.: N. Z. J. Agr. Res. *11*, 515 (1968).
247) Sullivan, J. M., Parker, M., Carbon, S. B.: J. Lab. Clin. Med. *71*, 893 (1968).
248) Mitchell, H. L.: J. Assoc. Offic. Anal. Chem. *52*, 222 (1969).
249) Dumanski, J., Waligora, B.: Zesz. Nauk Univ. Jagiellon., Pr. Chem. *13*, 185 (1968).
250) Munro, D. C.: Appl. Spectry *22*, 199 (1968).
251) Lyon, G. L., Brooks, R. R.: N. Z. J. Sci. *12*, 200 (1969).
252) Roth, F., Gilbert, E.: Mitt. Rebe Wein. Obstbau Fruechteverwert. (Klosterneuburg) *19*, 11 (1969).
253) Emmermann, R., Luecke, W.: Fresenius' Z. Anal. Chem. *248*, 325 (1969).
254) Hu, K. H., Friede, R. L.: J. Neurochem. *15*, 677 (1968).
255) Lukwakmann, C., McIntosh, J. E. A.: Nature *221*, 1111 (1969).
256) Kobes, R. D., Simpson, R. T., Valee, B. L., Rutter, W. J.: Biochemistry *8*, 585 (1969).
257) Gonick, P., Oberleus, D., Knechtges, T., Prasad, A. S.: Invest. Urol. *6*, 345 (1969).
258) Strehlow, C. D., Kneip, T. J.: Am. Ind. Hyg. Assoc. J. *30*, 372 (1969).
259) Ramirez-Munoz, J.: Anales Edafol. Agrobiol (Madrid) *26*, 1211 (1967).

260) Brian, G. W.: J. Marine Biol. Assoc. U. K. *49*, 225 (1969).
261) Ambler, J. E., Brown, J. C.: Agron. J. *61*, 41 (1969).
262) Boawn, L. C., Rasmussen, P. E., Brown, J. W.: Agron J. *61*, 49 (1969).
263) Nayrot, J., Ravikovitch, S.: Soil Sci. *108*, 30 (1969).
264) Halim, A. H., Wasson, C. E., Ellis, R., Jr.: Agron. J. *60*, 267 (1968).
265) Rogers, G. R.: J. Assoc. Offic. Anal. Chem. *51*, 1042 (1968).
266) Richards, G. E.: Soil Sci. Soc. Am. Proc. *33*, 310 (1969).
267) Toribora, T. Y., Shields, C. P.: Am. Ind. Hyg. Assoc. J. *29*, 87 (1968).
268) Pyrih, R. Z., Bisque, R. E.: Econ. Geol. *64*, 825 (1969).
269) Neal, P.: J. Assoc. Offic. Anal. Chem. *52*, 875 (1969).
270) LaFlamme, Y.: At. Absorption Newsletter 7, 101 (1968).
271) Hoover, W. L., Reagor, J. C., Garner, J. C.: Assoc. Offic. Anal. Chem. *52*, 708 (1969).
272) Bionda, G., Ciurlo, R., Biino, L.: Ricera Sci. *37*, 665 (1967).
273) Jordan, J.: At. Absorption Newsletter *1*, 118 (1968).
274) McBride, C. H.: J. Assoc. Offic. Anal. Chem. *48*, 406 (1965).
275) – J. Assoc. Offic. Anal. Chem. *50*, 401 (1967).
276) Cundiff, R. H., Dobbins, J. T., Jr.: J. Assoc. Offic. Anal. Chem. *49*, 521 (1966).
277) Nixon, G. S.: Caries Res. *3*, 60 (1969).
278) Brunelle, R. L., Hoffman, C. M., Snow, K. B., Pro, M. J.: J. Assoc. Offic. Anal. Chem. *52*, 911 (1969).
279) Herrmann, R., Lang, W.: Nucl. Activ. Tech. Life Sci. Proc. Symp. Amsterdam *1967*, 247–265.
280) Rousselet, F., Girard, M. L.: Ann. Biol. Clin. *25*, 987 (1967).
281) Kovatsi, A. V.: Chim. Chronika (Athens, Greece) B *32*, 109 (1967).
282) Angel, C. R.: Proc. Univ. Mo. Annu. Conf. Trace Subst. Environ. Health, 1st, *1967*, 51–55.
283) Anacleto, M. H. C.: Rev. Port. Farm. *18*, 93 (1968).
284) Hankiewicz, J.: Postepy Hig. Med. Doswiadczalnej *23*, 465 (1969).
285) Reynolds, R. J.: World Med. Instr. 7, 10 (1969).
286) Skinner, J. M.: Proc. Soc. Anal. Chem. *6*, 131 (1969).
287) Wacker, W. E., Coombs, T. L.: Ann. Rev. Biochem. *38*, 539 (1969).
288) Westerlund-Helmerson, U.: At. Absorption Newsletter *5*, 97 (1966).
289) Bartels, H.: At. Absorption Newsletter *6*, 132 (1967).
290) Ezell, J. B., Jr.: At. Absorption Newsletter *6*, 84 (1967).
291) Gutsche, B., Bafi-Yeboa, M., Czernicki, B., Herrmann, R.: Deut. Tieraerztl. Wochenschr. *75*, 486 (1968).
292) Zaugg, W. S., Knox, R. J.: Anal. Biochem. *20*, 282 (1967).
293) Devoto, G.: Boll. Soc. Ital. Biol. Sper. *44*, 424 (1968).
294) Roe, D. A., Miller, P. S., Lutwak, L.: Anal. Biochem. *15*, 313 (1966).
295) Galindo, G. G., Appelt, H., Schalscha, E. B.: Soil Sci. Soc. Am. Proc. *33*, 974 (1969).
296) Gersonde, K.: Anal. Biochem. *25*, 459 (1968).
297) Suzuki, M., Coombs, T. L., Vallee, B. L.: Anal. Biochem. *32*, 106 (1969).
298) Fuwa, K., Vallee, B. L.: Anal. Chem. *41*, 188 (1969).
299) Bond, A. M., Willis, J. B.: Anal. Chem. *40*, 2087 (1968).
300) Hall, F. F., Peyton, G. A., Wilson, S. D.: Techn. Bull. Regist. Med. Technol. *39*, 89 (1969)
301) Potter, A. L., Ducay, E. D., MC Cready, R. M.: J. Assoc. Offic. Anal. Chem. *51*, 748 (1968).
302) Christian, G. D.: Anal. Letters *1*, 845 (1968).
303) – Feldman, F. J.: Anal. Chem. *43*, 611 (1971).

Received July 3, 1970

Fortschritte der chemischen Forschung
Topics in Current Chemistry

Herausgeber:
A. Davison · M. J. S. Dewar
K. Hafner · E. Heilbronner
U. Hofmann · K. Niedenzu
Kl. Schäfer · G. Wittig
Schriftleitung: F. Boschke

Band 17
W. Demtröder:
Laser Spectroscopy
With 16 fig. III,95 pages
1971. DM 28,–

Band 18
R. C. Bingham/
P. v. R. Schleyer:
Chemistry of Adamantanes
With 4 fig. III,102 pages
1971. DM 36,–

Band 19
L. Maier and G. Zon/
K. Mislow: The Chemistry
of Organophosphorus
Compounds I
With 11 fig.,III,94 pages
1971. DM 34,–

Band 20
H.J. Bestmann/
R. Zimmermann:
The Chemistry of Organo-
phosphorus Compounds II
With III,147 pages
(In German)
1971. DM 58,–

Band 21
L. Eberson/H. Schäfer:
Organic Electrochemistry
With 10 fig. III,182 pages
1971. DM 58,–

Band 22
W. Kutzelnigg/G. Del Re/
G. Berthier:
σ and π Electrons
in Organic Compounds
With 8 fig. III,122 pages
1971. DM 48,–

Band 23
M.J.S. Dewar and
W.B. England/L.S. Salmon/
K. Ruedenberg:
Molecular Orbitals
With 40 fig and 5 tables
III,123 pages
1971. DM 32,–

Band 24
H. Fischer and J.F.Labarre/
F. Crasnier: Electronic
Structure of Organic
Compounds
With 12 fig. III,54 pages
1971. DM 18,–

Band 25
J. Manassen, R.L. Banks,
W. Strohmeier,
G.-M.Schwab,F.Steinbach:
Catalysis
With 26 fig. III, 154 pages
1972. DM 48, –

Springer-Verlag
Berlin
Heidelberg
New York
München · London
Paris · Tokyo · Sydney

In kritischen Übersichten werden in dieser Reihe Stand und Entwicklung aktueller chemischer Forschungsgebiete beschrieben. Sie wendet sich an alle Chemiker in Forschung und Industrie, die am Fortschritt ihrer Wissenschaft teilhaben wollen.

In der Regel werden nur Beiträge veröffentlicht, die ausdrücklich angefordert worden sind. Schriftleitung und Herausgeber sind aber für ergänzende Anregungen und Hinweise jederzeit dankbar. Manuskripte können in den „Fortschritten der chemichen Forschung" in Deutsch oder Englisch veröffentlicht werden.

Jeder Band der Reihe ist einzeln käuflich.

This series presents critical reviews of the present position and future trends in modern chemical research. It is addressed to all research and industrial chemists who wish to keep abreast of advances in their subject.

As a rule, contributions are specially commissioned. The editors and publishers will, however, always be pleased to receive suggestions and supplementary information. Papers are accepted for "Topics in Current Chemistry" in either German or English.

Any volume of the series may be purchased separately.